Developing Mathematical and Scientific Literacy:

Effective Content Reading Practices

Developing Mathematical and Scientific Literacy:
Effective Content Reading Practices

by

David K. Pugalee

University of North Carolina Charlotte

Christopher-Gordon Publishers, Inc.
Norwood, Massachusetts

Copyright Acknowledgments

Christopher~Gordon Publishers, Inc.
Bridging Theory and Practice

1502 Providence Highway, Suite 12
Norwood, MA 02062

800-934-8322 • 781-762-5577
www.Christopher-Gordon.com

Printed in the United State of America
10 9 8 7 6 5 4 3 2 1   09 08 07 06

ISBN: 978-1-933760-09-4
Library of Congress Catalogue Number: 2006940275

Table of Contents

Introduction

"I think I understand stuff when I'm in class.
When I try to do homework I get confused
and then I try to get help from the book—but
none of it makes sense. I get frustrated and
give up. If I can't do better in math and science
then I'm going to fail for the year."

—Julian, age 14

Many students sitting in classrooms are struggling with poor reading skills. Their lack of strong literacy skills contributes to poor academic performance across all subject areas. Julian is typical of many of these students. They want to do well and make efforts to understand content from their classes. But like Julian, many students can't successfully maneuver textbooks, so they feel frustrated and give up. The problem is sometimes compounded in classes such as mathematics and science where there are few direct efforts to help students develop literacy skills. Several factors contribute to students' difficulty with literacy as it relates to learning mathematics and science from text. Students have difficulty with reading comprehension; mathematics and science texts are often more difficult to read, and there is generally a lack of explicit support in developing reading comprehension within these content areas.

Mathematics and science performance are inherently related to how students process and understand text. Despite a commonsense acceptance that reading is an integral part of the foundation to success in mathematics and science, there are few concerted efforts focusing on reading in these disciplines. Such efforts must provide substantive development of process skills that integrate experiences with text with prior knowledge and other information in producing relevant meaning. The necessary emphasis should go beyond a traditional view of reading that is characterized by decoding and literal comprehension. A more comprehensive view of reading is necessary. The idea of reading literacy captures this broader, more comprehensive framework. Reading literacy includes understanding, reflecting, and using written information for multiple purposes (Organization for Economic Co-operation and Development, 2003).[1]

This idea of reading literacy recognizes the importance of a set of linguistic tools that allows the user to respond to multiple roles involving text. This book explores the development of reading literacy within the context of learning mathematics and science. In order to maximize students' potential for developing mathematical and scientific literacy, instruction must consider how to support students as they encounter mathematics and science text. There must be a focus on the role of reading comprehension in mathematical and scientific achievement. The information in this work develops those crucial links between reading literacy and the development of mathematical and scientific literacy. The attainment of these multiple forms of literacy is imperative if students are to be served through their educational experiences in the mathematics and science classroom. Literacy involves more than reading and writing. It encompasses comprehending what is read, reflecting and evaluating what is learned through texts, becoming engaged in the reading and writing processes, and knowing how to find and use knowledge in new situations (Wade & Moje, 2001).

Why This Book?

Why is this book so important? Students' reading performance continues to give reason for alarm. According to the 2003 National Assessment of Educational Progress (National Center for Education Statistics, 2004),[2] only 63% of fourth graders and 74% of eighth graders scored at or above the basic level on reading measures. The performance of minority and at-risk students was notably lower. International assessments also raise concern about students' reading performance. The Programme for International Student Assessment [PISA] is a survey of knowledge and skills of 15 year-olds in 30 industrialized countries that are members of the Organization for Economic Co-operation and Development. Results from the 2000 reading assessment (Kirsch et al., 2002) measured performance on a wide range of texts drawn from different situations involving three approaches or aspects to reading: retrieving information, interpreting, and reflecting (which includes evaluation). Means for students in the United States did not differ significantly from the average of the participating countries while scores from 12 countries (Australia, Austria, Belgium, Canada, Finland, Iceland, Ireland, Japan, Korea, New Zealand, Sweden, and United Kingdom) were significantly above the average. The national and international assessments of reading performance indicate a need for interventions that address the lackluster performance of US students.

There is an assumption that students in mathematics and science classes possess sufficient skills in using language, including reading comprehension and writing. This may especially be the case in reform-oriented classrooms that stress the development of reasoning and thinking. An emphasis on reading in mathematics and science provides hope for many students who are not being reached by current educational reforms. Readers construct meaning as their prior knowledge and experiences interact with texts. It is within this complex interaction that students working with mathematics and science texts develop crucial cognitive skills that allow them to construct meaning. Students need explicit support in how to develop these cognitive skills within the context of studying mathematics and science. While the process of developing reading literacy in mathematics and science is similar to other subjects, there is a need to emphasize the different nature of the texts in these subjects and to focus on content-area implementation of appropriate strategies. The absence of concerted efforts to address reading difficulties in mathematics and science provides a beckoned call for possible solutions. Teachers need information on how to help students become better readers of mathematics and science. This involves understanding the nature of the texts and appropriate strategies to help students read for understanding—to develop reading literacy in mathematics and science.

In addition, the current emphasis on high-stakes assessment necessitates supports so that students' reading skills do not confound performance on assessments in mathematics and science. Reading comprehension is by nature part of what is assessed in any written measure of performance. Today's mathematics assessments are characterized by contextualized problems that require students to understand and process text. Many of the traditional computation problems, what some educators refer to as "naked computation," have been replaced by "word-problems." These problems require students to use appropriate procedures and compute accurately, but they also require students to read and comprehend the situation in which the information is embedded. Science assessments frequently require students to read critically in order to answer questions about concepts and processes.

There is a direct relationship between reading performance and performance on mathematics and science tasks. The relationship between reading and a variety of academic measures is well established in the research literature (Gee & Rakow, 1990; Espin & Deno, 1993) as is the importance of reading as an important predictor of achievement (Roeschl-Heils et al., 2003). Further, these studies suggest that academic performance in content areas and reading performance measures can be positively impacted when teachers emphasize reading skills. Low reading achievement is identified as a primary factor in chronically low-performing schools (Moats, 1999). Flick and Lederman (2002) point out that reading comprehension is the goal of reading and that reading comprehension requires higher-level thinking to infer meaning, consider implications, and make decisions about applications. Scientific inquiry and mathematical problem solving embody these higher-level thinking processes. Students cannot be expected to do well if they cannot understand the texts they encounter in schools.

Effective mathematics and science instruction has high levels of engagement with literacy. Borasi and Siegel (2000) identify critical goals for focusing on reading in mathematics. These goals can be extended to include science content as well. This emphasis on literacy in mathematics and science can help students (1) construct and negotiate understanding of important ideas and concepts and (2) engage successfully in inquiry and problem solving. The attainment of these goals is essential if students are to develop the learning skills that are necessary to be mathematically and scientifically literate.

Organization of the Book

This book is organized in four chapters. Each chapter is described in the following paragraphs. The information is organized to provide a basic foundation in understanding the reading process and how it relates to student learning with particular consideration for the needs of mathematics and science teachers.

Chapter 1: Understanding the Development of Reading Skills and the Process of Reading

This chapter emphasizes the development of reading skills that promote comprehension and understanding. The primary goal of this chapter is to help mathematics and science teachers understand some basic foundational ideas about reading without getting bogged down in theory. Reading is a complex linguistic and cognitive skill that must be acquired. Teachers who have a working knowledge of some of the basic ideas related to reading development are better prepared to address literacy development as part of content area instruction. An understanding of the process of reading will increase instructional effectiveness in ways that help students read as a tool for content area learning.

Chapter 2: The Nature of Mathematics and Science Texts

This chapter provides an overview of how texts and approaches to texts in mathematics and science differ from other content areas. One goal will be to help teachers understand structure of texts in these disciplines and how these structures impact instructional decisions—so that they feel comfortable with their role as a teacher of reading.

Chapter 3: Teaching Reading in Math and Science

This chapter aligns instructional goals with literacy (reading) goals. This chapter provides a framework that will help teachers consistently consider the role of reading in supporting students' learning. Assessment will be introduced as an integral component of this process. One goal will be to help teachers develop a sound approach to mathematics and science instruction by considering literacy issues and skills along with their instructional goals.

Chapter 4: Strategies for Comprehension

The three reading processes measured in PISA assessments will serve as an organizational tool for this chapter. The emphasis will be on providing reading strategies that support understanding mathematics and science. One goal is to help teachers have a comprehensive set of strategies that support various reading processes while assisting them in understanding how to incorporate basic assessment procedures so that they can monitor the effectiveness of the reading strategies.

The three purposes for reading (as identified in the PISA assessments) are: retrieving information, interpreting texts, and reflecting to link knowledge with experience. *Retrieving information* involves the ability to locate information in texts. Readers should be able to retrieve information that is found explicitly in text. Such actions are often motivated by a question related to a task or to develop ideas and obtain information about some facet of the text. The focus of these reading processes typically involves a sentence or one or more blocks of text. Several pieces of information may be necessary to adequately respond to the question or task. Tasks include identifying information that is relevant to the reader's goal, looking for a specific idea, searching for definitions, identifying key concepts, and so on.

Interpreting information involves the ability to construct meaning and draw inferences from written information. Interpreting information involves the reader in building a general understanding of the text. The reader will be able to identify key components of the text including the main idea and the overall purpose of the text. Interpretation involves organization of ideas in the text to develop logical understanding of ideas and concepts.

Reflecting and evaluation involves the ability to relate text to their other knowledge, ideas, and experiences. Reflecting and evaluating requires the reader to connect information within the text as well as to outside sources of text and experiences. Supporting evidence, concepts, and ideas from the text are used in various levels of reasoning. The reader will be able to draw upon information from the text, including text features and language, to evaluate the quality and appropriateness of ideas, outcomes, and inferences.

Notes to Introduction

1. The Organization for Economic Co-operation and Development (OECD) is a collaboration of 30 member-countries that are committed to democratic government and market economy. Their work has a global reach involving issues related to macroeconomics, trade, education, development, and science and innovation. The Programme for International Student Achievement (PISA) is an OECD collaborative effort to measure how well students at age 15 are prepared to meet the challenges of today's society. The PISA assesses domain-specific knowledge in reading, mathematical, and scientific literacy. Reading literacy is the conceptual foundation for the literacy assessment. More information can be obtained online from the OECD (http://www.oecd.com).

2. The National Assessment of Educational Progress is sometimes referred to as the Nations Report Card. It is the only national and ongoing assessment of what students in the United States can do in various subject areas. Since 1969, these assessments have covered

reading, mathematics, science, writing, US history, civics, geography, and the arts. Information summarizes the performance of students in grades 4, 8, and 12 for the nation and geographic regions of the country including both public and nonpublic schools. The 2005 assessment will cover mathematics, reading, and science. The project is under the auspices of the U.S. Department of Education. Additional information is available at http://nces.ed.gov/nationsreportcard/.

Chapter
One

Understanding the
Development of Reading Skills
and the Process of Reading

Background

Reading and literacy has become a hotbed of public debate and the focus of educational policy. We'll not engage in that debate—which has been stormy at times—as the goal of this work is not to forward a position on the acquisition of reading skills, but to provide a source for supporting students' reading comprehension in mathematics and science. It is, nevertheless, important to recognize the importance of all of these perspectives and how they have contributed to the complex and interactive reading landscape in our schools today. This focus on reading, in and outside of educational and policy circles, underscores the importance of reading in producing an informed and educated electorate. Yet, many do not understand the complexity of reading as a process. Educators are not exempt from holding misconceptions about reading. For many years, reading was considered the instructional domain of the elementary teacher and rarely did content area teachers seriously concern themselves with addressing reading difficulties though they may have recognized that their students were having problems.

An understanding of the complex process that we call reading will help us understand that the job is too monumental to not be an integral part of every teacher's mission. It is important to have some understanding of the reading process in order to make informed decisions about content area literacy in mathematics and science. Teachers are influenced by their beliefs and knowledge. Understanding basic reading processes will provide background information that will influence how instructional decisions about teaching and learning mathematics and science. This chapter will provide some background information on reading that will help us understand the complexity of the reading process. An overview of the historical development of reading will be a good starting point for building this foundation.

A Historical Perspective on Reading

Historically, there are four periods of reading pedagogy (Turbill, 2002). In the age of decoding (1950s to early 1970s) there was an emphasis on learning the sound-symbol relationships and word recognition. Decoding might be viewed in terms of comparing familiar words to unfamiliar ones. A student who knows and recognizes the word science may use those skills to figure out the word scientist. Decoding involves an emphasis on the automatic recognition of single words. Many recall the endless phonics worksheets and basal readers that predominated reading instruction during this period. This skills-based approach to reading instruction promoted teaching of highly sequential phonics skills, typically in isolation, without attention to the development of meaning for those words. This is not to say that phonological skills are not important but to point out the focus of this period and the instructional emphases. This approach involved a lot of rote memorization and drill and practice.

Reading as meaning making took form in the early to mid 1970s. Sound-symbol relationships along with background knowledge were viewed as important in reading for meaning. This approach emphasized a lot of reading and being read to. This movement challenged the dominance of basal readers and the more traditional instructional approaches that supported their implementation. Advocates argued that basal approaches separated language and learning in artificial ways that made learning to read more difficult (Graves et al., 2004). Developmental reading programs became prominent as schools sought to address the problem of students reading below grade level. Perspectives from cognitive psychologists became a major influence during the 1970s prompting an emphasis on comprehension and a move beyond psychometric measures of reading to assessments that included identifying main ideas, describing important details, determining the sequence of events, comparing and contrasting, recognizing cause-effect relations, and drawing conclusions. The influences of this movement still shape reading policy and practices today.

The age of reading-writing connections took root in the early to mid 1980s. Learning sound-symbol relationships, background knowledge the reader brings to reading, reading for meaning, an emphasis on examining written models, understanding writing, and the writing process characterized reading pedagogy during this time. This approach argued for language-rich environments where individuals could engage in authentic literacy tasks. Emergent literacy proponents questioned conventional literacy practices arguing that literacy development begins before formal instruction in reading. This movement built on the momentum of the reading for meaning emphasis of the 1970s linking the meaning making processes of reading to those of writing. This connection was likely inevitable given that experts in writing were advancing ideas of writing as a process and writing-across-the-curriculum projects were becoming popular components of literacy programs.

Reading for social purpose began in the early 1990s. Dramatic and rapid social and educational changes underscored the need for higher levels of literacy. Literacy became a broad construct that involved the integration of speaking, listening, and critical thinking with reading and writing. There was an emphasis on the development of literacy in a range of contexts. This broader definition of literacy stressed an awareness of culture and class as they related to literacy. In addition, globalization and political influences focused attention on literacy. Politicization of reading led to policies and mandates related to reading instruction and reading performance.

Turbill (2002) described a fifth age that has begun to take shape in reading. This age of multiliteracies extends beyond reading text to include color, sound, movement, and visual representation. This new age is sure to include ideas and concepts related to information and technological literacies. The growth of media based documents will necessitate increased consideration of the skills necessary to function successfully in making meaning from these multiple forms of text. The abilities to assess the need for additional information and the type(s) of information necessary, to locate information from multiple sources in multiple formats, to evaluate the authenticity and value of multiple forms of information, and to synthesize and use this information are likely to be central tenets of this emerging perspective.

The Reading Process

What is reading? The answer to that question may seem obvious, but our ideas about what reading means are deeply entrenched in our philosophies that are constructed from years of personal experiences and observations.

These philosophies drive our literacy decisions and practices. It is important to reflect on our own ideas about reading so that we can better understand how reading and content area learning are interrelated. The legal definition of reading found in the No Child Left Behind Act of 2001 captures the complex and multifaceted nature of the reading process. The Act defines reading as (see Section 1208):

A complex system of deriving meaning from print that requires all of the following:

- skills and knowledge to understand how phonemes or speech sounds are connected to print,
- the ability to decode unfamiliar words,
- the ability to read fluently,
- sufficient background information and vocabulary to foster reading comprehension,
- the development of active appropriate strategies to construct meaning from print, and
- the development and maintenance of a motivation to read.

Regardless of our position in the reading debate, most could agree that this definition of reading is rather comprehensive. It is clear that reading is a complex and interactive process that involves the effective application of numerous cognitive skills. Luke and Freebody (1999) describe these skills for contemporary readers as code breaker, meaning maker, text user, and text critic. This is a complex set of skills that involves the active engagement of the reader. Reading places a lot of cognitive demand on students. The meaning an individual forms from text involves the interaction of the knowledge and skills of the reader and the message conveyed through the text (Graves et al., 2004). The skills and knowledge that readers bring to their interaction with text cannot be ignored. They play a pivotal role in the interplay with meaning created from experiences with text. This view of the complex and interactive nature of reading conflicts with the transmission model of teaching mathematics and science that is predominant in many classrooms. An interactive model suggests that teachers can structure instructional events and environments to support the development of literacy skills.

In general, reading can be envisioned of consisting of a two-part process (Freitag, 1997). The first involves transfer of encoded information from the text to the reader—the act of decoding while the second involves comprehension of the information. Decoding in its simplest terms involves the process readers use to figure out words. This may involve identifying sounds in words such as the sound of tri in triangle. Once the sounds are identified and these sounds are combined to determine the pronunciation, the reader knows the word. At a more basic level, readers must also understand the letters that represent words—often referred to as alphabetics. Successful readers need to know how smaller sounds make up words—phonemic awareness—and how letters and letter combinations represent these sounds—phonics and word analysis. Reading comprehension might be thought of as involving a complex string of subprocesses that work in conjunction with each other. Though the string provides a somewhat linear picture of the process, reading comprehension is much more complex than what can be captured in this description:

- perception of letters,
- rapid recognition of words,
- detection of function and meaning of words within a sentence,
- integration of parts of a sentence and sentences into meaning (Aarnoutse & Schellings, 2003).

The reader must be able to rapidly recognize strings of words as they process phrases, sentences, and blocks of text—fluent reading. Fluent reading is essential to reading comprehension (Kruidenier, 2002). For students to become independent readers, they must possess good decoding skills. Readers who have difficulty understanding key words will have difficulty comprehending the text.

Mathematics and science involves decoding many technical and concept-dense words. Developing a basic mathematics and science vocabulary and decoding skills is important for students to comprehend new words. As students add new words to their vocabulary, they are also adding to their ability to decode new words that they will encounter in reading mathematics and science texts. Instructional strategies to support students in decoding mathematics and science texts appear in the following sections.

The second process of reading involves comprehension of the information. The goal of reading instruction is to develop readers' comprehension skills. Comprehension involves understanding text and deriving meaning of key concepts. The inability to recognize key words or vocabulary is especially important in comprehending concept-dense texts such as those in mathematics and science. Students can have difficulty with any aspect of reading and this difficulty impacts their subject matter learning. Draper (2002) points out that experience and environment play a pivotal role in how students learn and that language plays a key role in the acquisition of knowledge. Comprehension depends on content knowledge of the reader and their ability to make sense of the signs and symbols in the text. These views of learning underscore the significance of attending to literacy skills in teaching content area concepts and skills.

Comprehension is at the heart of the development of mathematical and scientific literacy. It involves students in actively interacting with various texts to obtain information and create meaning. Reading comprehension involves several subprocesses that operate together as the reader makes meaning from text. These subprocesses include identifying a purpose for reading, activating prior knowledge, developing and maintaining motivation, and applying metacognitive skills. Each of these is discussed in the following section.

Fundamental Components of Successful Reading

When we think about what makes a student a good reader, many images come to mind. Many authors have described some of the attributes that "good" readers seem to possess. These characteristics provide some thinking points about how to build sound instructional programs that develop these attributes. What are the characteristics of good readers? Table 1.1 compares the characteristics of good and poor readers (Cook, 1989; Fuentes, 1998). A careful consideration of these characteristics will reveal that they align with the four subprocesses that work together to make reading comprehension possible.

Table 1.1. Characteristics of Good and Poor Readers

Characteristics of Good Readers	Characteristics of Poor Readers
• Activate prior knowledge • Understand purpose for reading • Choose appropriate strategies • Focus attention • Anticipate and predict • Use context when approaches new terms • Know text structure • Organize new information for integration • Self-regulate understanding and comprehension • Reflect after reading • Summarize major ideas • Seek outside information	• Begin reading without considering knowledge and experiences • Read without a goal • Use limited knowledge of how to approach text • Are easily distracted • Read quickly without reflecting • Skip important new vocabulary • Do not recognize text organization • Do not connect new information • Do not engage in self-monitoring • Stop processing when finished with text • Do not recognize or identify major ideas • Fail to use outside resources

Purpose of the Text

Students should identify a purpose for reading before they engage with the text. For one thing, the purpose of reading determines the strategies that will be employed to meet the goal or purpose for reading. A clear purpose for reading also plays an important role in activating prior knowledge the reader may have about the topic. The purpose for reading helps the reader select approaches to the text. Methods or approaches to reading include:

- Skimming: reading to gain a general understanding of the text
- Scanning: reading to locate specific information or details
- Reading the entire text (sometimes with re-reading): reading for clarification or to develop understanding of complex ideas or concepts
- Reading the entire text: reading for critiquing and evaluating the text; reading to get a general idea of details such as reading for pleasure

Readers who are successful know why they are reading the text and how the information relates to a task or goal. Students approach reading tasks differently. Identifying a purpose for processing text facilitates how students approach and implement learning strategies to meet the goals of the task.

Good readers are able to identify a purpose of the text. Identifying and understanding the purpose of a passage of text is a primary factor in reading for comprehension (Flick & Lederman, 2002). Students might read texts to study, to compose a response, to organize ideas for critical analysis or reflection, to identify information as quickly as possible, or for pleasure. Students who understand how their task involves reading text are more likely to be successful. Students should be able to adapt strategies to respond to multiple purposes for processing text. This adaptation process involves metacognitive skills, which are discussed later in this chapter. Recall from the Introduction that PISA (OECD, 2003) identifies three major purposes for reading: retrieving information, interpreting texts, and reflecting to link knowledge with experience. These purposes, which incorporate the purposes presented here, will provide a structure for discussing reading strategies later in this book.

Prior Knowledge

The level of understanding and previous knowledge a student possesses plays a critical role in how they assimilate new information into their existing knowledge structures. Students who have large gaps in knowledge or possess no relevant background related to a concept will have great difficulty understanding the new information. This most generally results in great frustration and the student constructing no real new knowledge or understanding. In order for learning to occur, students need to incorporate their new understanding into what they already know. This explains, to a degree, what happens when students know something for a test or quiz but then fail to know the information in other contexts at a later time.

Prior knowledge is often referred to in cognitive terms as schema. If someone doesn't understand something, we may say that person lacks a schema for that concept. Schemata involve one's knowledge about things, situations, experiences, and the sequences of events and actions. Schema not only involves an individual's knowledge about the world but includes knowledge that is related to particular subject-area content. Understanding the schemata that students bring to learning situations in mathematics and science is important for planning effective instruction. Students who approach complex concepts and situations in mathematics and science depend on a whole body of knowledge about those ideas as well as a body of knowledge about how to organize and execute a problem-solving plan. This complex level of information requires facility in activating knowledge so that information fits within a developing cognitive system. Additionally, students need to engage in self-monitoring so that schemata are used effectively to develop mathematical and scientific understanding from text. This self-monitoring interrelationship, or metacognition, is discussed later.

Schema is very important to development of conceptual understanding. Consider the examples of schema in mathematics and science that appear in Table 1.2. Students in the mathematics example are beginning a lesson on the Pythagorean Theorem. In order for a student to understand the theorem and be able to apply it, they must have some understanding of geometry, algebra, and number concepts. For example, the student will need to understand triangles and recognize that right triangles (triangles with a 90° angle) have special properties. Students might think about how theorems come about and the historical nature of mathematics including who is Pythagoras. Students will also need to have background knowledge about solving multi-step equations including working with the squares and square roots of numbers. Students' schema will also include information about their experiences working with triangles and right triangles. Of course, students will have varying complexity of the schemas they have related to the goals of this lesson:

1. Identify the parts of a right triangle.
   a. Identify the right angle in a right triangle.
   b. Identify the sides of a triangle including the base and hypotenuse.
2. Use the Pythagorean Theorem to solve problems.
   a. Find the missing length of a side given the lengths of two legs.
   b. Solve problems that involve creating a model using the Pythagorean Theorem.

Students in the science lesson are studying a lesson on the nature of sound. Students' background will include their understanding of waves. Their understanding might be limited—such as the waves of an ocean. Vibrations are an important part of understanding the nature of sound. Students' schema may include information and experiences with vibrations. Knowledge and facts about the ear and the role it plays in hearing and the production of sound are also important. Sounds also have different levels. Students will possess varying levels of knowledge about these various concepts and some students may not have any background relative to a particular component. All background knowledge and experiences will shape how students come to understand the following lesson goals:

1. Explain the nature and properties of sound
   a. Compare and Contrast how sound travels through air, water, and solids.
   b. Illustrate how sound is made by vibrations.
   c. Describe changes in the quality of sound, including pitch and echoes.
2. Identify and describe uses of sound.
   a. Describe how sounds are used.
   b. Create or modify devices that produce sound and describe their purpose and effect.

Table 1.2.   Schemas in Mathematics and Science

Mathematics Example	Science Example
Context: Introductory lesson on the Pythagorean Theorem (middle grades)	Context: Introductory lesson on sound (elementary)
Schema: • Triangles, including relationship of lengths of sides, right triangles • Numbers that are squared and square roots • Solving multi-step equations • Historical context of mathematics • Theorems and their use in mathematics • Possible real-world examples of applications • Strategies for approaching text	Schema: • Waves and their general characteristics • Vibrations and how they are produced • General knowledge based on experiences with sound • Relationship of ears to sound and hearing • Sounds have different levels • Strategies for approaching text

Engaging and Motivating the Reader

When readers are engaged they read to understand, enjoy learning, and have confidence in their reading abilities (Guthrie, 2001). Motivated readers are behaving deliberately and purposively during reading (Guthrie & Wigfield, 1999). Teachers provide classroom contexts for engagement by promoting a coherent approach to reading including instruction in reading strategies. Guthrie describes a sound research base that shows engaged readers comprehend text and have higher levels of reading achievement (also see Roeschl-Heils et al., 2003). Engaged readers are motivated, strategic, knowledgeable, and socially interactive. Evidence suggests that learners who develop strong self-concepts related to their cognitive abilities perform better. Research substantiates a strong connection between an individual's feelings about reading or reading attitude and reading comprehension (Aarnoutse & Schnellings, 2003).

Motivational variables are difficult to specify and describe. Interest in reading is related to motivation to read. Motivation is considered to be a factor in activating and guiding reading behavior (Aarnoutse & Schellings, 2003). Interest in reading should be encouraged. One of the components that play a part in developing an interest to read is success in reading. Teachers can promote an interest in reading by using a variety of texts in various instructional activities. Some ideas for motivating readers are presented in Chapter 4.

It would seem rather obvious that there is a positive relationship between students' interest in reading and reading comprehension. A bi-directional relationship is reported between reading motivation and the application of reading strategies. Reading motivation influences the use of reading strategies and the use of reading strategies influences reading motivation. One way to improve students' self-concepts is to positively impact the development of metacognitive skills (Pape & Smith, 2002).

Metacognition

We have seen that reading components involve a complex interplay between cognitive actions of the reader. This complex interplay involves metacognition. Metacognition involves "knowing about knowing." More formally it is the act of self-monitoring and the application of learning strategies. It is not an automatic process but emerges as a result of development of cognitive systems of thinking and reasoning (see Jacobson, 1998). Metacognition also includes beliefs and attitudes about learning and one's capabilities in various learning situations. Successful readers develop a level of expertise that allows them to predict their performance relative to a given task. Metacognition includes knowledge of resources as well as control of one's thinking (Pugalee, 2005). Metacognitive variables are the best predictor of reading performance (Roeschl-Harris et al., 2003).

In reading, metacognition involves the individual's ability to regulate their learning by reflecting on the task. Metacognitive behaviors associated with reading include self-knowledge, task knowledge, and strategy knowledge (Graves et al., 2004). Metacognitive skills are especially important when the reader encounters a word, phrase, or idea that interrupts the reading process (Underwood, 1997). The reader must know what to do to figure out the information and return to the reading task. This process of figuring out what to do involves the selection of methods or approaches that will provide success in meeting the learning targets or goals. Metacognitive skills can be taught by focusing on developing self-awareness and self-assessment skills.

Metacognition influences reading on several key variables: texts, tasks, strategies, and learner characteristics (Ambruster, 1983; Collins, 1994). The features of text influence students' reading approaches and ultimately their comprehension. Developing an awareness of text structures and their impact on reading can have an impact on comprehension. Self-regulated readers analyze tasks and set goals to successfully respond to the task, monitor and control their actions during implementation, and make judgments relative to their progress and alter their approaches accordingly (Pape & Smith, 2002). These might be referred to as planning, regulation, and evaluation. Successful readers are able to monitor their approaches and adjust to facilitate comprehension in ways that make the processing of text a smooth process. Students who have less developed metacognitive skills tend to focus on words and decoding and lack knowledge of strategies and approaches to assist when encountering difficult or unknown words or information. Adjusting reading rates, using context and experience to identify unknown text, mapping, summarizing, and self-questioning are among strategies students might employ. Strong metacognitive skill involves not only having knowledge of these strategies and their effectiveness but a reader's awareness of their own background, skills, and experiences and how these are likely to impact reading. It is important for students to self-reflect and analyze their approaches and performance so that they have an accurate sense of their abilities and which strategies are effective in addressing their limitations. Metacognition also involves knowledge about the use of general strategies for remembering and recalling information. The central theme of metacognitive development is to raise awareness. Awareness of strategies, awareness of how to implement strategies successfully, and an awareness of how those strategies are working are pivotal in developing students' reading capabilities.

Reading and Sign/Symbol Systems

A discussion of reading processes in mathematics and science would not be complete without some mention of the importance of sign and symbol systems that are used freely in these disciplines. In both mathematics and science symbols and signs are used to express concepts in very concise and exact terms. For example, a middle grades science unit dealing with work and energy freely uses the symbol g to denote free fall acceleration, which is 9.8 m/s^2. Chemistry also relies heavily on communicating ideas and concepts through symbols. For example the statement $2Mg + O_2 \longrightarrow 2MgO + \text{heat}$ means that magnesium metal reacts with oxygen to produce magnesium oxide and heat. Then there are symbols for poison, biohazard, flammable material, corrosive material, and so on. Elementary students also encounter weather symbols as well as diagrams and models, which represent multiple concepts. In mathematics, $x \in N$ *means* that x belongs to the set of natural numbers or the set of numbers 1, 2, 3, 4, …. The Σ means summation and Δ is used to signify change. Consider that $4(2+4)^2/(3!)$ means to find the sum of 2 and 4, square the result and multiply by 4. Take this result and divide it by 3 factorial—that is, 3 times 2 times 1 which is 6. Did you get 24? Then there are additional symbols used in geometry and in the study of logic.

Semiotics is the field of study that deals with the concepts of sign systems and their meaning. These experts believe that a sign is the basic unit of meaning that is common to all such systems and can be anything that represents something to an individual (Berghoff, 1998). The sign itself does not transmit meaning to the reader. Seeing or hearing a word or sign does not automatically result in the reader or hearer constructing a meaning for that sign. Berghoff provides an example of a reader encountering the word *chien* but not immediately having a concept of dog—unless the reader speaks French since *chien* is the French word for dog. Symbols used in mathematics and science also possess these characteristics which make it difficult for readers to understand the frequently technical and complex nature of the meanings behind those signs.

Berghoff (1993) describes how we use sign systems in coordinated ways. The use of maps to explain spoken directions, the blending of art and words by authors and illustrators, the use of gestures to emphasize spoken language, and the supplementing of spoken language by gestures, postures, glances, grimaces, shrugs, and grunts. As such, sign systems allow us to know and express that knowledge in multiple ways that are complementary.

Adams (2003) offers that reading mathematics requires understanding the terminology used in the discipline; symbols and numerals used in context; the relationship between words, numerals, and symbols; and the order of words, symbols, and numerals. Mathematics and science depends on symbol systems and the use of graphs and images to convey rather abstract notions and concepts. Students experience difficulty reading such representations with ease. Given their common use in mathematics and science texts, reading instruction must focus on how to assist students in approaching such text structures so that comprehension of the signs and symbols contributes to the overall comprehension of the concepts and ideas presented in the text.

Some Ideas on Assessment

One of the questions that arises in considering how to support mathematics and science teachers in providing effective interventions for reading is the issue of assessment. Assessing students' reading performance can be an overwhelming prospect particularly for mathematics and science teachers who may not have experience or background to help them identify and address problems. While there are numerous formal means of assessing reading performance, these forms of assessment frequently do not provide readily usable information to guide teachers in making instructional decisions that will help students be more successful in comprehending mathematics and science texts. Assessment provides the teacher with a better understanding of the reading process and students' knowledge of mathematics or science. The focus on assessment for teachers of mathematics and science isn't on discrete skills but on the process of reading. Inherent in this approach to assessment is an emphasis on comprehension. Assessment of reading in mathematics and science can be viewed in holistic terms as the assessment of comprehension processes. The International Reading Association (IRA, 2000) asserts that assessment should identify strengths as well as needs of students and involves students in making decisions about their learning.

Our discussion of the characteristics of good readers can be synthesized into a model that moves us toward a more explicit consideration of assessment. Arrasmith and Dwyer (2001) have developed a trait-based model of an effective reader that complements their recognized 6+1 Trait writing assessment model.[1] Their reading model identifies six traits of an effective reader:

- conventions
- context
- synthesis
- comprehension
- interpretations
- evaluation

The developers reason that a trait-based assessment of reading is effective because the system defines effectiveness for complex reading tasks, produces a continuum for considering reading performance, supplies a common language for characterizing the quality of reading, and encourages self-reflection and analysis of students. This system of thinking can provide a useful framework for mathematics and science teachers to consider students' reading levels.

Table 1.3 provides an adaptation of the characteristics of these six traits recognizing the expository nature of mathematics and science texts. These modifications include symbols and models as important features of text. Another important modification is an emphasis on reasoning over the construction of interpretations and opinions recognizing that facts and information guide students to formulate conclusions that must be scientifically and mathematically based. Notice the hierarchical nature of these traits. Teachers can use these characteristics as a way of determining where students are in terms of their reading capabilities but also as a tool for determining the level at which instructional activities focus. This framework provides an informal means of assessing students' reading performance.

Table 1.3. Characteristic Reading Behaviors for Six Traits

Traits	Characteristic Reading Behaviors
Conventions	• Decodes writing standards of word recognition, punctuation, and grammar; decodes symbols, graphics, and models • Decodes organization standards of the text • Recognizes the genre • Reads with fluency and expression
Comprehension	• Determines a purpose for reading • Uses appropriate reading strategies • Identifies main ideas, key concepts, sequence of events • Identifies significant and supporting details • Recognizes the meaning conveyed through symbols, graphics, and models • Summarizes text
Context	• Determines the goal or purpose of the author • Understands that reader's point of view and background affects understanding • Sees the bigger picture that comprises or connects concepts • Identifies applications of concepts and ideas
Interpretation	• Identifies problems, incongruencies, or missing information in text • Analyzes text for information that addresses problems, incongruencies, or missing information • Modifies interpretation based on information, data, evidence, or outcomes • Relates analysis to bigger picture • Creates arguments based on ideas and concepts not included in the specific text
Synthesis	• Uses comparison and contrast strategies • Incorporates background knowledge and experiences • Structures information for explanation and analysis • Determines cause and effect relationships and/or describes models that allow predictions based on analysis of relationships • Identifies and compares multiple sources in order to analyze text
Evaluation	• Formulates conclusions and reasons about the text • Formulates questions related to the text • Distinguishes fact from opinion, essential from unnecessary conditions • Uses reasoning and proof to substantiate judgments and conclusions

It is also important for students to become active in evaluating their own reading success. The traits framework could be modified with simpler language to use with students. The following are some general questions that students could answer to provide some informal assessment about their reading abilities (Jacobs & Paris, 1989):

(1) Do you ever go back and read text over and over again?

(2) Do you ever stop while you are reading and try to predict what will occur next?

(3) Do you use strategies such as underlining, taking notes, or visualizing while you are reading?

(4) Do you think about what sentences mean and how they fit together?

(5) What do you do when you encounter a word you do not know?

(6) What do you do when you come to pieces of text that you do not understand?

(7) What do you do when you read to prepare for a test or other task?

These assessment strategies provide teachers with tools that allow a continuous link between teaching and assessing. Formal assessments do not provide the type of data that teachers need to make instructional decisions that impact individual student learning on a day-by-day basis. The assessment strategies presented here provide an effective and efficient mechanism for teachers on multiple components of reading processes.

Summary

Understanding some basic information about the reading process provides an essential foundation to guide instructional decision making. This chapter provided an overview of some basic principles of reading. Decoding, reading as meaning making, reading-writing connections, and reading for social purpose were offered as four periods of reading history. A fifth emerging period, called multiple literacies, emphasizes that reading includes color, sound, movement, and visual representation. Reading perspectives paint a picture of a complex cognitive activity that involves decoding and comprehension processes. Comprehension involves several subprocesses that influence the reading success of students. Identifying a purpose of the text, prior knowledge, engagement and motivation, and metacognition interact through a cognitively complex interplay as the reader encounters text. In addition, symbol and symbol systems present special challenges for reading instruction in mathematics and science. Teachers can informally assess students' reading performance and abilities by focusing on characteristics related to six traits of effective reading: conventions, context, synthesis, comprehension, interpretations, and evaluation.

Notes to Chapter 1

1. The 6+1 Trait model provides a framework that uses characteristics of writing to create a common vision of "good" writing. The six traits are ideas, organization, voice, word choice, fluency, convention and the +1 component is presentation. The goal of the framework is to provide a tool for assessing writing while also improving writing skills and positively affecting instruction. Additional information is available at the Northwest Regional Educational Laboratory web site at http://www.nwrel.org. The reading model was developed to complement the assessment process used for writing. Complete rubrics and assessment information are available through NWREL.

Chapter Two

The Nature of Mathematics and Science Texts

In order for students in mathematics and science to fully participate in their own learning, they must be able to access and understand texts that provide information and exchange ideas and concepts. Students may find that their prior experiences with text do not always map nicely in dealing with mathematics and science texts. Students frequently have no real strategies or plans for how they will approach using these texts to guide their learning. Understanding these features is essential for the teacher so that she/he can guide students in developing successful reading approaches. Students also benefit from understanding these structures so that they can use that knowledge to develop plans and select strategies that will help them to effectively read the text with understanding. This section of the book will discuss some of the features of texts in mathematics and science. Before beginning that discussion, let's explore some basic underpinnings relative to the importance of text structure knowledge on reading performance.

When we think of learning mathematics or science, we sooner or later get back to the importance of language. The printed word presents its on challenges in this complex interplay between language and mathematics or science learning. Reading mathematics and science texts is related to metalinguistic awareness. Students need to develop metalinguistic awareness which affects their ability to reflect on and analyze a language and how it works. Metalinguistic awareness supports the reader as he/she reflects on the structure and functional features of a text, make choices about how to communicate that information, and manipulates the units of language (MacGregror & Price, 1999). For example, a student who has not developed good metalinguistic awareness may miss out on understanding words with multiple meanings, words used as metaphors, or analogies. These actions are inherent to the study of mathematics and science and can be facilitated by providing instruction and support for students as they navigate mathematics and science texts. Metalinguistic awareness of text structures is important in promoting reading comprehension even at primary grade levels (Garner & Bochna, 2004). Interestingly, research shows that instruction about text structure not only improves reading comprehension but also positively affects development of writing process skills (Ambruster, Anderson, & Ostertag, 1989).

Textbook Use in Mathematics and Science

Mathematics and science depend on the use of textbooks as a primary source for teaching and learning content. One estimate is that more than 75% of classroom time and 90% of homework time involve textbooks (Woodward & Elliott, 1990) and more than 70% of mathematics and science teachers depend primarily on the use of textbook-based curricula (Weiss, Pasley, Smith, Banilower, & Heck, 2003). Textbooks frequently are used for verifying facts, accessing diagrams or illustrations, and to study for tests. Students may rarely initiate references to textbooks which may signal that students do not have a working knowledge about how to use the features of textbooks to guide and support their own learning. While teachers may often direct students to parts of a text, students rarely use their texts as a primary source for content information.

Teachers may feel frustrated by the large size of textbooks and some think that they are responsible for covering all the material (Valverde & Schmidt, 1997/98). This creates an apprehension based on two general assumptions (McTighe, Seif, & Wiggins, 2004). First, teachers should "cover" material and if the information is discussed and assignments given then students will remember it. Second, standards should be addressed one at a time through planning for lessons. McTighe and his colleagues argue for "uncovering" content by focusing more on fewer topics with more depth. It is clear that part of the solution for in-depth understanding should include an emphasis on how comprehension and understanding are facilitated through students' and teachers' use of texts. The second assumption underscores the importance of planning as part of effective instruction. This reiterates the need for lessons carefully aligned to learning targets (often communicated through state curricular frameworks) and particular attention to how reading and comprehending text are related to the goals of the lesson. Such an approach, as described earlier in this book, assists teachers in understanding standards and identifying how multiple learning targets can be accomplished in one lesson.

Students do very little content reading, neither in mathematics or science nor any content area, in class or outside (Wade & Moje, 2001). Several reasons are offered for this lack of interaction with texts. Students find classroom texts to be difficult to comprehend and not very engaging, teachers do not often assign textbook readings as a primary source of information, and many teachers rely on explanations, lectures, notes, demonstration, and other direct teaching methods to deliver course content. Students come to realize that they can rely extensively on the oral texts presented in class, including class notes and worksheets, to learn the material and respond to related learning tasks. Textbook-based instruction frequently proceeds under the assumption that students can read and develop meaning from their interaction with the text (Bryant, Ugel, & Hamff, 1999).

Reading in the early grades focuses more on narrative texts that are generally student friendly. By fourth grade, students face substantive increases in the demands of reading: academics focuse more on the textbook; vocabulary is less conversational and familiar but more technical and specialized; abstract ideas are more prevalent; syntax of texts increases in complexity; and texts emphasize inferential thinking and prior knowledge (Allington, 2002).

Teachers should consider the difficulty level of texts when selecting materials for students so that the material is not too easy but also not too difficult. While mathematics and science teachers might not be comfortable employing readability formulas when considering text difficulty, there are some factors that could be considered in selecting texts that are appropriate for students (Graves, Juel, & Graves, 2004, pp. 252–253):

- Familiarity of content
- Required background knowledge
- Organization
- Unity of writing
- Quality and verve (enthusiasm or "gusto") of writing
- Interestingness
- Sentence complexity
- Vocabulary
- Length

Students who are aware of the organization of texts understand how ideas and information are arranged and how that information is related through the structure of the text. We saw earlier how metacognitive skills are used in helping readers identify and implement various approaches to comprehending text. Informational texts employ the use of less familiar text patterns such as compare and contrast, and cause and effect (Brown, 2003) that are different from familiar narrative structures. Knowledge of text structures plays a significant role in students' successful selection of strategies.

Types of Text

Mathematical and scientific texts are diverse and extend beyond traditional conceptions of traditional textbooks. Even within textbooks, there are multiple text formats that serve important functions in communication of information and ideas. OECD (2003) identifies text as either continuous or non-continuous. Continuous texts include sentences and paragraphs and are probably what most of us think of when we think of text. Non-continuous texts may be categorized by formal structure. Such formats include charts and graphs, tables and matrices, diagrams, maps, forms (formatted texts that require readers to respond), information sheets, advertisements or calls, vouchers (such as tickets or invoices), and certificates (various forms of agreements or contracts). These text formats often require students to understand how information is represented and organized so that it is useful in understanding the technical ideas that are the common function of these text types.

Mathematics and science texts depend on the use of non-continuous features to present conceptually dense information in concise formats. Figure 2.1 is an example of non-continuous text depicting the relationship between age and the number of sit-ups. This example from the National Assessment of Educational Progress

Figure 2.1. Relationship between Age and Number of Sit-ups

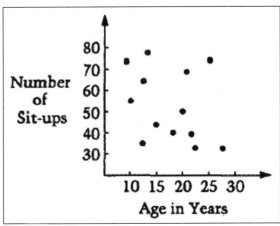

(NAEP) asked eighth graders (item was a common question repeated in the twelfth grade mathematics assessment) to find the median of the data (National Center for Education Statistics [NCES], 2001).

Research shows that the majority of students do not use or have limited benefit from pictures, diagrams, tables, charts, and other forms of non-continuous text (Novak, Mintzes, & Wandersee, 2000). Wandersee (2000) describes some ways that students can be supported to use photographs as a learning tool: writing descriptions of the images; exchanging those written descriptions with peers to uncover and eliminate missed features; and orally interpreting the images in terms of content. These ideas can be extended to include various forms of images, illustrations, and diagrams in both mathematics and science.

Mathematics and science texts include various forms of print media such as textbooks, worksheets, newspapers; encyclopedias, journals, manuals, and computer-oriented information; and so on. These texts are primarily informational texts presented in prose format and sometimes document format as in the use of charts, tables, diagrams, and so on. Though young children get little exposure to or training to read informational texts, they comprise the majority of comprehension assessments (Brown, 2003). In the NAEP (NCES, 2001) test item below, the nature of such assessment is evident. Two sample questions used on the 1998 grade 4 reading assessment are also included to demonstrate how informational texts are used in assessment of reading performance.

Figure 2.2. NAEP Fourth Grade Reading Passage

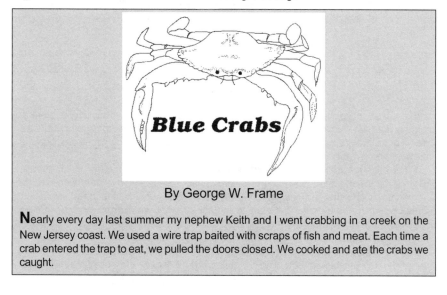

By George W. Frame

Nearly every day last summer my nephew Keith and I went crabbing in a creek on the New Jersey coast. We used a wire trap baited with scraps of fish and meat. Each time a crab entered the trap to eat, we pulled the doors closed. We cooked and ate the crabs we caught.

Figure 2.2. NAEP Fourth Grade Reading Passage *(Continued)*

Blue crabs are very strong. Their big claws can make a painful pinch. When cornered, the crabs boldly defend themselves. They wave their outstretched claws and are fast and ready to fight. Keith and I had to be very careful to avoid having our fingers pinched.

Crabs are **arthropods**, a very large group of animals that have an external skeleton and jointed legs. Other kinds of arthropods are insects, spiders, and centipedes. Blue crabs belong to a particular arthropod group called **crustaceans**. Crustaceans are abundant in the ocean, just as insects are on land.

The blue crab's hard shell is a strong armor. But the armor must be cast off from time to time so the crab can grow bigger. Getting rid of its shell is called **molting**.

Each blue crab molts about twenty times during its life. Just before molting, a new soft shell forms under the hard outer shell. Then the outer shell splits apart, and the crab backs out. This leaves the crab with a soft, wrinkled, outer covering. The body increases in size by absorbing water, stretching the soft shell to a much larger size. The crab hides for a few hours until its new shell has hardened.

Keith and I sometimes found these soft-shell crabs clinging to pilings and hiding beneath seaweed.

Blue crabs mate when the female undergoes her last molt and still has a soft shell. The male courts her by dancing from side to side while holding his claws outstretched. He then

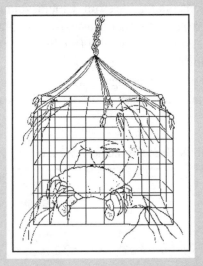

transfers sperm to the female, where they are stored until egg laying begins several months later. The female blue crab mates only once but receives enough sperm to fertilize all the eggs that she will lay in her lifetime. Usually she lays eggs two or three times during the summer, and then she dies.

When the eggs are fertilized and laid, they become glued to long hairs on the underside of the female's abdomen. The egg mass sometimes looks like an orange-brown sponge and contains up to two million eggs until they hatch — about nine to fourteen days later. Only one of the blue crabs that we caught last summer was carrying eggs, and we returned her to the water so her eggs could hatch. Most females with eggs stay in the deeper, saltier water at the ocean's edge rather than in the marshes.

The young blue crabs, and most other young crustaceans, hatch into larvae that look very different from their parents. The tiny blue crab babies are hardly bigger than a speck

Figure 2.2. NAEP Fourth Grade Reading Passage *(Continued)*

of dust. They are transparent and look like they are all head and tail. These larvae swim near the surface of the sea, and grow a new and bigger shell every few days. They soon change in shape so that they can either swim or crawl around on the bottom. Then they molt again and look like tiny adult crabs. After that their appearance does not change, but they continue to molt every twenty or thirty days as they grow.

As blue crabs become older, some move into shallower waters. The males in particular go into creeks and marshes, sometimes all the way to the freshwater streams and rivers. Keith and I caught ninety-two blue crabs in the shallow creek of the tide marsh last summer. Eighty-seven of those crabs were males, and only five were females.

Gulls find and eat many blue crabs. They easily catch crabs that hide in puddles at low tide. Other predators are raccoons, alligators, and people. If caught, the crabs sometimes drop off a leg or claw to escape. Seven of the blue crabs that Keith and I caught were missing a claw.

Crabs are able to replace their lost limbs. If a leg or claw is seriously injured, the crab drops it off. The opening that is left near the body closes to prevent the loss of blood. Soon a new limb begins growing at the break. The next time the crab molts, the tiny limb's covering is cast off, too, and the crab then has a new usable leg or claw. The new limb is smaller that the lost one. But by the time the crab molts two or three more times, the new leg or claw will be normal size.

Many fishermen catch crabs to sell. Most are caught in wire traps or with baited lines during the summer while the crabs are active. In the winter, the fishermen drag big nets through the mud for the dormant crabs. Commercial fishermen catch a lot of crabs, sometimes more than 50 million pounds in a year. And many other crabs are caught by weekend fishermen who crab for fun and food.

The blue crab has a scientific name, just like all other living things. Its name is *Callinectes sapidus*. In the Latin language *Callinectes* means "beautiful swimmer," and *sapidus* means "delicious." I think that scientists gave the blue crab a very appropriate name.

Last updated 8 November 2001 (VYG)

3. The growth of a blue crab larva into a full-grown blue crab is most like the development of
   - A)   a human baby into a teen-ager
   - B)   an egg into a chicken
   - C)   a tadpole into a frog
   - D)   a seed into a tree

4. Write a paragraph telling the major things you learned about blue crabs.

A futuristic look at texts must consider multimedia formats. There are various forms of text available on the Internet and other platforms such as CD-ROMS. Other formats include electronic texts, e-books, and various digital formats that are growing in availability while becoming more affordable. The multimedia formats present opportunities to provide sound and animation to supplement text. Such features provide promise for making information more accessible to readers; however, developers and authors should consider basic ideas related to the reading process so that the texts present ideas and concepts in clear and engaging ways.

Characteristics of Mathematics and Science Texts

Mathematics and science texts have some common features. Texts use signaling devices such as headings and subheadings to chunk key concepts. These devices can help students identify key concepts and approach reading sections of text with an organization scheme in mind. Such thinking helps students create meaningful mind-maps for storing information. Typographical features such as use of special fonts, boldface, or italics words, is another common feature that says "pay attention" to this particular word or phrase. These words or phrases usually involve a key concept or a term that is essential to understanding the text. Graphics and illustrations are also common in both mathematics and science. They relate to the text in specific and meaningful ways—sometimes clarifying or elaborating on key concepts. Diagrams and illustrations also provide examples and tools for students to organize information being presented in the text. These features often require students to read right to left, up or down, and sometimes diagonally or in other directions instead of left to right as they are accustomed to doing with narrative texts (Barton, Heidema, & Jordan, 2002). Glossaries, tables of contents, indexes, and review questions are additional navigational tools that are used effectively by adept readers (Brown, 2003).

Mathematics texts

Mathematics texts generally have some common features. Understanding these features and helping students understand and use them, are important in approaching reading tasks. Mathematics texts do not characteristically contain a lot of text to read, but the seemingly small amount of text carries very conceptually dense information, sometimes in five to eight short sentences. Schell (cited in Barton, Heidema, & Jordan, 2002) reports that mathematics texts might contain more concepts per line, sentence, and paragraph than other kinds of texts. The text may be divided into sections that include diagrams, cases or examples, models, or illustrations that are used to clarify or broaden the conceptual information being presented. Any figures or diagrams emphasize the ideas presented through text and are fundamental to developing a solid understanding of the information and concept or concepts being discussed. Often these features are interspersed with definitions, rules, or formulas. Examples are presented before students are given additional problems to apply the concepts covered in the material. It is important to note that these text features differ dramatically from those in other content area texts thus presenting challenges for students to fully comprehend the information (Pugalee, 2005).

Wakefield (2000) identifies several characteristics of mathematics that underscore the complex nature of the words, numerals, and symbols that readers of such texts must understand. These characteristics include:

- Abstractions such as verbal and written symbols representing ideas and images
- Uniform and consistent use of symbols and rules
- Linear and serial expressions
- Practice increases understanding
- Memorization of symbols and rules is important for success
- Translations and interpretations are required at different levels for novice and advanced learners
- Symbol order influences meaning
- Encoding and decoding is required for effective communication
- Fluency is marked by intuition and insightfulness
- Experiences are the foundation for future development
- There are infinite possibilities for expressions (adapted, pp. 272–273).

The complexity of dealing with the specialized language of mathematics and the resulting difficulty in reading comprehension is emphasized in the following table (Thompson & Rubenstein, 2000). The table presents an abbreviated version of their work. These examples demonstrate how important it is for mathematics teachers to explicitly support students as they encounter the specialized language of mathematics and science.

Table 2.1.   Vocabulary and Language Issues in Mathematics Texts

Category of Issue	Example
Some words common in everyday English have distinct mathematical meanings.	Number: power Geometry: leg Discrete Math: tree
Some words common in everyday English have comparable mathematical meanings but are more precise.	Algebra: limit Geometry: simila Statistics: average
Some mathematical terms are only encountered in mathematical contexts.	Number: decimal Algebra: integer Statistics: permutation
Some words have multiple meanings.	Geometry: round Statistics: range Discrete Math: inverse
Modifiers may change meanings.	Algebra: root or square root Geometry: polygon or regular polygon Statistics: number or random number
Some mathematical phrases must be understood as a unit.	Number: at most Algebra: one-to-one Geometry: if-and-only-if
Some words also used in science have different technical meanings.	Algebra: solution Statistics: experiment Discrete Math: element
Some math terms sound like everyday words.	Number: sum or some Geometry: theorem or theory Discrete Math: complement or compliment
Some math terms are related and students confuse meanings.	Number: numerator and denominator Algebra: solve and simplify Discrete Math: converse and inverse
Technology may use specialized language.	Number: EXP for scientific notation Algebra: LOG for base-ten logarithms Statistics: LinReg for linear regression

Science texts

Science texts are characterized by several features that increase students' comprehension difficulties:

- proportion of topics written in argumentative and expository forms;
- proportion of statements that are expressed as true, likely to be true, of uncertain status, likely to be false or false;

- frequency of metalanguage including observation, method, conclusions, and so on;
- the proportion reflecting scientific status of and roles in scientific reasoning (Penney, Norris, Phillips, & Clark, 2003).

These texts utilize definitions, classifications, illustrations, comparison and contrast, and functional and causal analysis that increase the conceptual density of the material. Science texts present a large number of concepts with vocabulary a major emphasis in the material. The emphasis on vocabulary may sidetrack students from focusing on the big picture—the major scientific concepts—and connecting them in relevant and meaningful ways (Grossen & Romance, 1994). The technical language of science has evolved in order to classify, decompose, and explain utilizing scientific genres such as reporting, explaining, and experimenting (Halliday & Martin, 1993).

Science texts use a lot of different features to provide supporting information related to the concepts and ideas. Research notes that children do not react positively to these features and more often are confused by the overload. The illustrations, graphics, and highlighted items require instruction on how to use these as part of making sense of the texts. It is noted that some children approach reading these texts as other types of text, going from word to work and from top to bottom. The frequent jumps to sidebars or extra-text features serve to confuse students. Other students may look at pictures and captions but never read the lesson titles so the information is not appropriately chunked or assimilated into a working schema (Budiansky, 2001; Roseman, Kulm, & Shuttleworth, 2001).

Some of the common organizational or layout features of science texts include multiple headings and subheadings to delineate treatment of various concepts or ideas; multiple graphics including pictures, charts, tables, and illustrations reinforcing and expanding information presented in the text; vocabulary text blocks often appearing at the beginning of a chapter or section; sidebars with questions and ideas including supplementary information intended to elaborate or extend the information presented; summary information; questions or exercise.

Using Text Features Effectively

Some simple exercises can provide information for teachers about how students use the text features while also providing opportunities for explicit instruction to help students understand how those features support learning key content. The following examples are derived from Ruddell's (2005) approach to evaluating instructional materials. In using these exercises, it might be helpful to contextualize them in specific lessons that students are currently considering. It is not necessary to do all of them at one time but as they occur in the texts. Review of the features may be necessary to help students continue to be aware of how text features signal the organization of ideas and information.

Table 2.1.   Exercises to Help Students Understand Text Features

1. How can you tell examples from explanations? Give an example from a lesson in the text.
2. What does the highlighting signal in the text? Provide an example from the text and tell how the highlighting helps you understand the lesson.
3. Where do you find rules, definitions, theorems, or generalizations in the text? Provide an example of two of these from the text.
4. How would you use the text to help you complete a homework exercise? Be specific and provide details (listing steps may help you think about what you would do).
5. What are the benefits of the summaries or conclusions sections?
6. How would you interpret this figure or diagram? How does your interpretation relate to the text?
7. Where do you go in the textbook when you do not know the meaning of a word?
8. What steps are necessary to complete this investigation or problem?
9. How do the headings in the text help you organize information? Give an example from the text.
10. How do the pictures or illustrations relate to the text material? Give one example and tell how it assists you in reading the text.

Activities and discussion that help students make important conceptual connections between their own learning, text features and organization, and the reading process are vital in content area instruction. Experiences should help students (1) develop connections between what they know and the information in the text, (2) monitor the adequacy of the meaning(s) they construct while engaging with the text, (3) repair faulty comprehension while reading, (4) distinguish important from less important information in the text, (5) synthesize information within the text and across multiple information sources, (6) draw inferences during and after reading text, and (7) asks questions to guide comprehension as they read (Ruddell & Ruddell, 1995).

Schumm, Vaughn, and Saumell (1994) identify several teacher practices for modifying texts. Some of these include previewing the textbook with students after determining their reading levels relative to the textbook difficulty; providing direct assistance including read alouds, individual or small group instruction, and cooperative learning groups; simplifying the text through the use of highlighting or color codes and constructing abridged versions of text or text passages; supplementing the text with computer programs, films, and other mutli-level and multi-material approaches; adapting teaching methods to enhance student learning of text information through use of summaries, outlines, questions to guide reading, views of assignments, vocabulary review or introduction, assisting with answering text questions, guided class discussion, and other pre, during, and post reading activities; and providing direct instruction on reading strategies that can be used independently by students such as modeling effective use of comprehension strategies, instruction on use of graphic aids, or teaching memory strategies.

Conclusion

The conceptual density of mathematics and science texts is marked by a high dependency on prior knowledge, the technical language, the patterns of argumentation and fact stating, and the use of canons of evidence. This requires the use of explicit strategies to help students construct meaning from these texts particularly since the majority (and in some cases all) of their experiences with reading instruction has focused on narrative genres. Various strategies that emphasize summaries of information, conceptual mapping, and thematic organizers provide opportunities for readers to comprehend information and to understand how texts are arranged.

In order to be effective, students need support in identifying and using reading strategies that assess the importance of information and prior knowledge; set a purpose; summarize; infer meanings; monitor comprehension; utilize text structure; reason critically; improve memory; self-regulate to address comprehension problems; and skim, elaborate, and sequence (Penney, Norris, Phillips, & Clark, 2003).

Chapter Three

Teaching Reading in Math and Science

Literacy Development and the Teaching of Mathematics and Science

Learning the language of mathematics and science remains a barrier for many students. Reading texts in these disciplines in a primary source of the language that students' encounter in their pursuit of developing mathematical and scientific literacy. Goals of mathematics and science instruction include, as a central tenet, the development of reasoning. The nature of the texts that students encounter incorporates this instructional target through information, concepts, and approaches that are language rich. The landscape of the mathematics and science classroom must explicitly address the role of reading texts in this construction of understanding.

Teachers of mathematics and science can effectively approach reading instruction within the contexts of content teaching by considering the interaction between the text and the learner. The goal of this chapter is not to engage in the debate on reading models but to provide some guidelines that can be applied in mathematics and science classrooms. The interactive approach combines components from several appropriate models. This approach considers the experiences and knowledge that the learner brings to text situations, builds comprehension through recognizing ideas that the learner has about the meaning of the text, emphasizes the importance of text, and recognizes the importance of the reader using all levels of information to make meaning from the text. This approach emphasizes the interaction between the reader, the environment, and the text. As such, the model provides a comprehensive and workable approach for those who do not have formal training as teachers of reading.

This chapter provides ideas and principles about teaching reading in mathematics and science. The discussion begins with a framework for instructional planning, continues with discussion of key components that affects students' reading comprehension. Various issues and processes related to providing literacy instruction are presented ending with a model that provides a process for providing strategy instruction to support reading comprehension.

Literacy and Teaching Mathematics and Science

Expanding the role of reading in mathematics and science instruction complements an emphasis on teaching for understanding. Reading supports should not be viewed as an add-on to instructional practices but should be integrated into classroom routines. Borasi and Siegel (2000) caution against an overemphasis on reading instruction that focuses on helping students comprehend word problems or learn technical vocabulary. These are important instructional goals but comprehensive reading programs extend reading instruction to include multiple purposes and text types. Borasi and Siegel argue that the design of meaningful experiences that integrate reading, writing, and talking provide rich opportunities for learning and that the development of reading practices should become part of a class's everyday events that promote learning communities that value collaboration and meaning making.

Effective mathematics and science teaching supports students in organizing their thinking about complex and related concepts and phenomena. Effective teaching also builds students' acquisition of the language of mathematics and science that allows the communication of ideas and concepts that support individual learning and public discourse. In order to meet these broad educational goals, mathematics and science teaching will need to incorporate sound literacy instruction as an integral part of what it means to engage in mathematical and scientific inquiry.

These ideas stress that teaching is driven by beliefs and experiences and teaching reading in the content is subject to the philosophy of teaching and learning. Reading instruction must be guided by a larger philosophical frame that supports effective mathematics and science instruction. Emphasizing this philosophical framework is essential so that teachers are driven by the same guiding principles that inform their instructional decisions for content teaching. If the importance of language and reading are part of this belief system then reading instruction will not be perceived as another add-on to an already crowded curriculum. Stephens and Brown (2005) assert that content literacy is "neither an add-on to the curriculum nor is it a substitute for content. It provides teachers and students alike with effective tools for learning the content of any subject" (p. 8).

What are some guiding principles of a philosophy of teaching and learning mathematics and science that recognizes the power of language, and the necessity of reading instruction, in promoting students' development of mathematical and scientific literacy? Ball and Bass (2003) assert that mathematics is characterized by a compression of information into forms that are both abstract and usable. This level of compression is efficient for mathematicians; however, teachers need the skills to "decompress" or "unpack" these conceptually dense ideas. The National Science Education Standards (National Research Council, 1995) relates this complexity of ideas while also underscoring the importance of multiple forms of communication. An important stage of inquiry and learning science involves oral and written discourse that focuses attention on what students know and how their knowledge connects to larger ideas, other domains, and the world beyond the classroom. Teachers must promote different forms of communication (including spoken, written, pictorial, graphic, mathematical, and electronic) in order to guide student discourse. In order for students to be successful in these language-intense environments, teachers must provide opportunities for the development of language skills including reading. Reading provides the link for integrating contextual information with prior knowledge in an active construction of meaning. This is facilitated by students' reading comprehension and other communicative abilities. Students must be able to read text critically and engage in discourse about content related concepts, ideas, and processes. This level of thinking and reasoning will not be sufficiently developed if there is not a focus on the relationship between scientific and mathematical thinking and reasoning and the processing of text.

Effective Instructional Planning

Effective instructional planning includes consideration about the role of reading as students engage in a lesson. Support for reading should not be viewed as "something else to do." Appropriate strategies assist students in meeting the desired mathematics and science learning outcomes or objectives. Helping students read successfully is essential in developing comprehension and understanding. A planning framework (Tovani, 2004) provides a useful tool for focusing attention on learning outcomes and for considering key components that address the learning and reading link. This framework includes five central elements:

1. Instructional Purpose (Essential Understanding for Students)
2. Difficult Concepts/Ideas
3. Approaches for Negotiating Difficult Concepts/Ideas
4. Outcomes from Reading
5. Reading Supports

This framework doesn't supplant the instructional design process. It provides a means of focusing on how to effectively teach mathematics and science concepts while considering the role that text might play in that particular lesson. The framework is easily integrated into the thinking that teachers do as they plan for instruction.

The first component addresses a primary concern of mathematics and science teachers as they deliberate about how to address reading issues as they teach content. The lesson centers on what is essential for students to know as a result of instruction. Teachers consider local, state, and national goals in defining the key concepts or ideas that are central to the lesson. This means that teachers are focusing on teaching mathematical and scientific content. The second and third components ask teachers to anticipate where students might have the most trouble and to identify possible approaches for helping students negotiate the material. Teachers might think of models or strategies that help students understand key ideas and concepts. The fourth and fifth components focus more explicitly on reading. It is important to understand what students should be able to do as a result of their interactions with text. Wiggins and McTighe (2001) in their backward design process for instructional design encourages the consideration of outcomes before identifying experiences to help students meet those outcomes. This framework supports that approach by focusing on outcomes and then the identification of supports that will help students perform successfully. The fifth component asks for careful reflection on the types of reading supports that can help students as they develop and apply essential understanding and knowledge.

The following examples demonstrate the application of this framework. The examples are intended to be illustrative of the process. The texts that each lesson are based on are approximately 1½ pages in length excluding follow-up questions and homework. Notice how reading becomes a central part of each lesson. The use of strategies to support reading is a key component of the instructional framework. Discussions later in this chapter will focus on how to teach strategies to students. Strategies should always be used in the context of developing content. The specific strategies listed are described more fully later in this book.

Table 3.1. Instructional Planning Framework

	Mathematics	Science
Instructional Purpose (Essential Understanding for Students)	Students will be able to specify part-whole relationships.	Students will be able to describe why cells are the building blocks of life.
Difficult Concepts/Ideas	Function of denominator and numerator.	Cells are microscopic and have very specialized functions.
Approaches for Negotiating Difficult Concepts/Ideas	Use multiple representations: comparing items in a group; dividing a whole into parts (use circles/rectangles)	The WOW factor: how many cells are there in the human body? Relating the complexity of cell functions to the "building block" concept.
Outcomes from Reading	Students will identify key words denominator and numerator and be able to illustrate related concept.	Identifying key concepts and synthesizing those to support that cells are the building blocks of life.
Reading Supports	Vocabulary/Concept Strategy: students will identify terms and meaning including making appropriate illustrations; students will write a story depicting the relationship.	Concept Synthesis Strategy: students will identify several key concepts and then write a synthesis of those ideas.

General Ideas about Effective Reading Instruction

Before we begin a discussion that focuses on developing reading instruction, some basic tenets about effective reading instruction will provide a useful launching point. Richard Allington (2002) identified six common elements of effective literacy instruction that were found in research involving effective teachers: time, texts, teaching, talk, tasks, and testing. This discussion frames these important perspectives within content area literacy instruction in mathematics and science.

Time

In Allington's research, he found that activities that favor reading and writing were in greater proportion that other activities (remember that the focus was literacy instruction). Yet, time remains one of the important factors facing mathematics and science teachers when they consider implementing content area reading instruction into their subject area teaching. The bottom line is that ultimately the teacher's job is to provide instruction so that students learn mathematics and science; however, effective and efficient learning will be hampered if students do not have solid reading skills. Providing reading instruction and support within the instructional context of mathematics and science is a time efficient means to support student learning while honoring the commitment to provide the best possible teaching to help students develop content area knowledge and skills.

Texts

Allington posits that teachers who provide time for students to read so that they experience success stimulate motivation to read. Provide opportunities for multiple literacy-related events that emphasize texts other than the traditional textbooks. Newspaper articles, tradebooks, magazines, Internet sources, supplementary expository texts, and so on contain much information that can be used to emphasize mathematics and science while infusing instruction with texts that may be of higher interest for students. Teachers are encouraged to build a collection of mathematics and science related texts that will promote student engagement and content area learning.

Teaching

Allington suggests that teaching should involve active instruction including direct and explicit modeling of strategies to use when reading either in small groups, individually, or whole class settings. If the driving force behind teaching is to develop understanding then mathematics and science instruction will be characterized by active engagement that values literacy as an essential component of helping students learn subject area content.

Talk

Allington found exemplary teachers encouraged, modeled, and supported problem-posing and problem-solving talk in which concepts, hypotheses, and strategies are discussed. In fact, questioning is such an important component of effective instruction that we'll return to this topic later in this chapter along with a discussion of the link between reading-writing-talk as central to powerful learning environments. Cooperative learning, peer editing and review, and paired learning activities support effective communication in mathematics and science. A comprehensive approach to developing students' mathematical and scientific literacy involves an interactive model where reading, speaking, and writing interplay in a dynamic instructional environment (Pugalee, 2005).

Tasks

Allington found that successful teachers selected tasks that were challenging and required greater degrees of self-regulation. He included this as part of a "managed choice" program that allowed students some decisions about their work. Giving students choices also serves as an important motivational tool. Students will be motivated and challenged more from authentic reading and writing tasks than more traditional assignments (Mizell, 1997). Relevant tasks relate to students needs and have real-world connections. It is also important that tasks are varied, both in content and complexity, so that students develop a broad understanding of multiple mathematical and scientific topics, competencies, and skills while further refining reading skills appropriate for a wide range of texts requiring multiple reasoning approaches to promote comprehension and understanding (Pugalee, 2005).

Testing

Allington found that effective testing focused on improvement and not achievement. He advocates that teachers should know their students so they can grade with improvement evaluation as the goal. Assessment should include a variety of approaches that supports a variety of content-related measures and supports reading, writing, speaking, and listening.

These general ideas provide a beginning point to think about literacy instruction. We can agree that these ideas characterize some important considerations about teaching and learning. You will find these generalized principles embedded in the following discussion designed to provide more specific information to guide teachers in developing teaching practices that support reading—practices that are intrinsic to Allington's six "Ts" of effective literacy instruction.

Essential Teaching Practices

Evidence supports instructional methods that focus on causes of success and failure rather than praising performance or pointing out failures and methods that emphasize learning goals are more likely to promote students' confidence in their abilities (Hall et al., 1999). DiGisi and Yore (1992) offer several suggestions for helping students with reading comprehension in science. These suggestions are likewise applicable to mathematics (see Freitag, 1997). These suggestions include: using advance organizers to prepare for new information; identifying the author's purpose such as exposition, instruction, or practice; and employ a series of conceptual questions to assess students' learning. Related to identifying the purpose of text, Loranger (1999) adds that students should be taught to differentiate between narrative and expository texts. She also underscores the importance of activating prior knowledge and the role of metacognition in literacy development. A synthesis of these elements provides us with a comprehensive framework containing seven elements that are relevant and powerful for content area literacy.

1. Teaching students to differentiate between narrative and expository texts
2. Teaching students strategies for activating prior knowledge
3. Helping students understand how they learn (metacognition)
4. Teaching various strategies (Loranger, 1999)
5. Motivating readers
6. Questioning
7. Facilitating students construction of mental images
8. Reinforcing the reading-writing link (Dickinson & DiGisi, 1998).

Differentiating between Types of Text

As pointed out earlier, reading instruction primarily focuses on narrative text whereas science and mathematics texts are expository intended to provide detail about technical information and concepts for which the reader is unfamiliar. Chapter 2 focused on the nature of mathematics and science texts.

The information presented in that chapter provided ideas that extend beyond recognition of text as narrative or expository to focus on text features and the role they play in reading comprehension. As students are able to identify text types and features, they will be more capable of selecting strategies and approaches that will facilitate their understanding of the text with which they are engaged.

Activating Prior Knowledge

Students who activate prior knowledge about a concept or idea have a readily available mental framework on which to 'hang' or organize new information or ideas. This activation of prior knowledge is a key component in promoting effective comprehension of text. It is especially imperative for students to be aware of prior knowledge particularly when that knowledge is based on misconceptions or contains flaws or errors. As that prior understanding is made explicit, ideas and information that is incongruent with new information can be refined and reorganized.

Mathison (1989) includes the use of analogies as an effective means of promoting students' making connections to prior knowledge. She argues that analogical thinking helps students create critical schematic connections. For example, an analogy of a city's transportation networks provides a visual element for framing a lesson on keeping the heart healthy. Framing questions to help students think about how prior experience is related to a piece of text is a frequently used tool to help students consider related topics and ideas before engaging in learning new concepts or ideas. This process of reflecting and recording and/or discussing is effective in identifying what students already know about a topic. Active previewing of a piece of text is another way that teachers might help students make connections with prior knowledge or provide an opportunity through direct instruction to fill in gaps in background knowledge. In some cases teachers may need to provide explicit instruction to build background knowledge that will assist students in relating text information.

Metacognition

Recall from Chapter 2 the importance of metacognition on overall academic achievement as well as reading performance. As students become more self-reflective they become more capable of identifying their own learning characteristics including strengths and weaknesses. This awareness emerges from opportunities to self-assess and provides a foundation that builds in order for students to manage and regulate their own learning. Self-assessment of goals can also be extended into peer-assessments. Peer assessment can very effective in assessing students' writing if students understand the criteria that are to be applied (Pugalee, 2005). The same principles apply in peer assessment of the use of reading strategies. As students share their approaches and thinking related to a reading task, they are building a peer feedback loop that builds confidence, extends thinking and reasoning, and stimulates motivation. Teachers should assist students in thinking about their own learning, reflecting on the knowledge and skills that they possess, and show them how to use various appropriate strategies to facilitate reading and comprehending.

Metacognitive skills can be developed in classrooms where teachers make thinking about strategies and processes public. One way of accomplishing this visibility of cognitive processes is to model tasks providing details and descriptive procedures about approaching the task. This "thinking out loud" process can also be effectively incorporated by having students share their approaches and strategies discussing what worked well and any problems that arose during the reading. These instructional strategies not only work for developing metacognitive skills related to reading but are effective procedures for all content (Pugalee, 2005).

Some pedagogical practices that have a metacognitive emphasis include previewing material, activating prior knowledge, determining text characteristics, determining a purpose for reading, generating questions, predicting, verifying predictions, recognizing a comprehension breakdown, studying processes, trying to understand how students function, trying to get an underlying misconceptions, rereading, skimming, summarizing, paraphrasing, looking for important ideas, reading ahead for clarification, sequencing events, and relating new information to prior knowledge (Hall et al., 1999).

Teaching Various Strategies

There are two main emphases in content area teaching strategies: vocabulary development (including word recognition) and comprehension. Students must be able to recognize words quickly in order to promote both fluency and comprehension. Word recognition becomes especially important given the large and often complex vocabulary that students are introduced to in mathematics and science. Students who do not recognize words quickly often get bogged down in trying to figure out the word resulting in cognitive overload that prevents them from focusing on reading the text for meaning (see Bryant, Ugel, & Hamff, 1999). Word identification involves contextual, phonetic, and structural analysis. Most older students have developed decoding strategies such as breaking the words into syllables, identifying affixes and suffixes that carry particular meanings, and attending to root words and words with common meanings. These strategies address the structural and phonetic analysis in word recognition. Contextual analysis becomes an especially important skill to emphasize in mathematics and science because it involves using context clues such as charts, diagrams, and illustrations which are frequent features of texts in these disciplines.

Understanding key terms in mathematics and science involves more than just recognizing the words though word recognition is an important part of content area reading instruction. Vocabulary is rich in concepts and instruction in vocabulary development is essential to successful reading skills development and students' mathematical and scientific understanding. Graphic organizers are one way to support students in organizing vocabulary they encounter in texts. Graphic organizers represent a wide selection of tools that generally provide a visual representation to help students organize information. Reading strategies often employ various types of graphic organizers to support students' abilities to organize ideas and information.

Comprehension skills

Various forms of text are used to develop students' understanding of mathematics and science. One of these forms of texts is worksheets or activity books. While these have a place in reinforcing the development of concepts and skills, these forms of text are low level in both their cognitive and motivational demand and provide few opportunities for students to become reflective about their strategies or approaches (Hall et al., 1999). Considering comprehension at three levels provides a useful lens for thinking about cognitive engagement during reading: literal, interpretive, and applied. Literal comprehension is based on students' identification of information. The interpretive level involves drawing conclusions and inferences. Applied level comprehension requires considering information with a critical or evaluative eye and also involves relating information to prior experiences and background knowledge. These comprehension levels align with the reading processes of PISA (OECD, 2003): retrieving information, interpreting texts, and reflection and evaluation. These three reading processes will provide a framework for presentation of reading strategies later in the book. Webb (2002) provides descriptions of levels of knowledge related to reading, writing, mathematics, and science. These four levels of knowledge provide a framework for considering the depth of students' thinking relative to these core subjects. These descriptions give teachers a tool for differentiating students' levels of thinking.

Table 3.2.   Levels of Knowledge

	Reading	Writing	Mathematics	Science
Level 1	Refer to details in text; recall or simple understanding of a word or few words	Recite simple facts and basic ideas; listing, brainstorming; appropriate grammar/punctuation	Recall of information such as facts, definitions, terms, simple procedures; performing a simple algorithm or applying a simple formula	Recall and reproduction; recall information such as facts, definitions, simple procedures; performing a simple science process
Level 2	Intersentence analysis of inference; summarizing and predicting	Extemporaneous writing or drafts; connect ideas using basic organizational structure	Involves decisions on approach to a problem or activity; interpreting information	Involves decisions on approaches to a problem or question; making observations, collecting, displaying data
Level 3	Thinking beyond the text; Reasoning, generalizing, and connecting ideas	Composition development with multiple paragraphs; editing and revising; awareness of purpose and audience	Strategic thinking involving reasoning, planning, using evidence; describing thinking	Strategic thinking involving reasoning, planning, using evidence; drawing conclusions, citing evidence, creating arguments
Level 4	Apply information to new situation or task; hypothesizing and analyzing	Multiple paragraphs demonstrating synthesis and analysis of complex ideas or themes; deep awareness of purpose and audience	Complex reasoning, planning, developing; making several connections, relate ideas, select and defend alternatives	Make several connections, relate ideas within and among content areas; consider and defend approaches; experimental design and planning

Motivating the Reader

How can content area teachers in mathematics and science engage and motivate students as readers? One important component is to establish learning goals that clearly identify important targets. Second, promote students' self-efficacy about reading and other skills. Third, provide opportunities for students to engage in social learning communities. Guthrie (2001) identifies several instructional components that foster engagement and reading outcomes: coherence, conceptual orientation, real-world instruction, autonomy support, interesting texts, strategy instruction, collaboration, teacher involvement, rewards and praise, and evaluation.

Maitland (2000) identified several principles that have varying degrees of influence in motivating readers. These principles can be adapted to provide some challenging ideas for thinking about instruction in mathematics and science:

1. guide students to set personal goals;

2. provide some opportunities for students to select their own materials and activities;

3. allow students some sense of control coupled with encouragement and reinforcement from the teacher;

4. encourage students to set challenging tasks for themselves, self-assess their progress, and reflect on their accomplishments.

What do these principles imply for teachers of mathematics and science? These four principles have a lot to do with the role of the teacher in instruction. Teachers, through guidance and support, provide students with opportunities to take responsibility for their own learning thus removing them from passive roles as learners. Individualizing feedback so that students can focus on recommendations is vital in moving students toward a more responsible role. The teacher's role in this learner-centered environment is marked by facilitating, guiding, and modeling; by modeling approaches or strategies that are essential to learning; by tapping prior knowledge and providing students with schemata and scaffolding for concepts and processes; by providing opportunities for teacher/student and student/student interaction; and by developing an awareness of cognitive strategies that can be used to increase memory capacity (Winstead, 2004).

Allowing students' control of facets of their instruction such as selecting subtopics for projects and reading, materials for learning (books, electronic media, journals, magazines, etc.), and strategies that they believe are successful promote autonomy and motivate (Guthrie & Cox, 2001). Other identified means of motivating students include identifying and announcing a knowledge goal; providing a brief real-world connection related to the goal; teaching cognitive strategies for reading texts; assuring collaborative learning; and aligning assessment of student work to the context of the goal. In mathematics, for example, use student-centered and student-generated problems so students relate personally to material.

Pressley (1998) identifies five methods to improve reading motivation. The first involves positively impacting the attributions of poor readers by providing explicit instruction on reading strategies but also by addressing negative attributions. The second involves providing students with texts and books on topics that strongly interest them. Third, the classroom should be a rich reading and writing environment. Fourth, make a large number of books available in order to stimulate students' reading. Finally, classrooms should emphasize a community-of-learners approach that involves interaction and problem-oriented approaches.

Jacobs and Paris (1987) report that explaining thinking skills that could be used before, during, and after reading are beneficial to students of varying ages and reading ability levels. Note that this process meshes well with self-questioning, summarizing, paraphrasing, and predicting that characterize reciprocal teaching.

Questioning

Consideration of effective pedagogy in mathematics and science would be incomplete without explicit attention given to the role of questioning, particularly as questioning relates to helping students become better readers. Questioning is a critical teaching and learning tool that assists students in understanding subject matter content. Effective questioning should be extended to help students deal with text in mathematics and science. Such practices should become commonplace in the classroom as teachers and students pose and discuss questions that arise from interactions with text. Effective questions before reading provide a mechanism to help students activate prior knowledge, make predictions, and connect to big ideas not always explicitly referenced in the text. During reading, questions play a pivotal role in helping students make comparisons and generalizations, identify key concepts and ideas, and clarify meaning. As students reflect after reading, questions guide the location of details and information; recall of events, processes, or procedures; and summarizing of the major concepts. Teachers who help students formulate guiding questions promote comprehension of text while supporting the understanding of important mathematics and science concepts. Proficient readers become adept at using questions to guide their understanding of text.

Questions can help guide students' comprehension as they deal with text. Table 3.3 provides some key questions that teachers can use as part of mathematics and science instruction to help students make connections to prior knowledge, create visual images, draw inferences and conclusions, make judgments or interpretations, use main ideas to develop meaning, synthesize, monitor comprehension, explore author's intent, and use the structure of the text. There is substantive research showing that questioning is related to students' academic achievement. The use of higher order thinking questions is positively related to the development of students' literacy skills (Taylor, Pearson, Peterson, & Rodriguez, 2003), mathematics achievement (Balfanz & MacIver, 2000) and science performance (Wise, 1996). Some characteristics of effective questioning include focusing on salient content elements to improve comprehension; asking frequent questions to improve learning facts; posing questions before reading to improve interest; redirecting and probing when related to accuracy, clarity, and plausibility; and increasing wait time beyond three seconds (Cotton, 1988). Some student benefits from using higher-order questioning were also identified by Cotton. These include increasing on-task behavior, the length of student responses, the number of relevant student contributions, the frequency of student-to-student interactions, the use of complete sentences by students, the speculative thinking of students, and the number of relevant questions posed by students. Teachers should

Table 3.3.   Questions that Promote Comprehension

How does this connect with what I already know?	Activating relevant, prior knowledge before, during, and after reading
What pictures does this text create in my mind?	Creating visual and other sensory images from text during and after reading
How can I use the pictures and the text to help me understand?	Drawing inferences from the text to form conclusions, make critical judgments, and create unique interpretations
What are the most important ideas and themes in the text?	Using the main ideas to provide clues about meaning
How can I say this in my own words?	Synthesizing what they read
Does this make sense?	Monitoring comprehension
Why did the author write this?	Exploring the author's intent
How is this text like other texts that I have read?	Finding clues in the text's structure

From Ontario Ministry of Education. (2003). *Early Reading Strategy: The Report of the Expert Panel on Early Reading in Ontario*, 2003, p. 20. Toronto, Ontario: Queen's Printer for Ontario.

develop strong questioning skills that provide more opportunities for students to engage in higher level questions such as those that involve inferences, synthesis, and application. These are the types of questions that should be infused throughout mathematics and science instruction.

Assist Students in Constructing Mental Images

The construction of mental images is a particular strategy that has not received much attention in the literature dealing with teaching and learning reading, mathematics, and science. It involves deep mental structures that can be developed as part of students' reading strategies repertoire. Consider the role of imaging in learning important mathematics and science concepts. An important component of successful problem solving is the ability of students to develop a mental representation contextualizing the information and the application of a strategy or approach. In mathematics and science pictorial representation or images convey mathematical and scientific concepts. Such images are analogues of a mathematical or scientific concept designed to mirror the structure of an abstract concept (English, 1999). Many mathematical concepts, for example, involve symbolic, numerical, and graphic representations. Students who have developed some capacity to construct mental images are better able to make representations of mathematical situations. The same is true for development of mental images that involve representing scientific concepts.

In mathematics, for example, visual models are positively related to students' understanding of mathematics. Diezmann and English (2001) discuss the importance of diagrams, or visual representations that display information in a spatial layout, in helping students understand the structure of a problem. Diagrams become a visual tool for mathematical thinking and learning. Visual images, of which diagrams is a specialized case, are important to mathematics instruction. Several suggestions for developing students' expertise in using diagrams are adapted from the work of Diezmann and English and generalized to the development of students' use of various mental images.

1. The use of mental images can be supported through modeling and discussing their use.

2. Provide explicit instruction in the use of visual imaging.

3. Focus on the structural information represented in images rather than surface features.

4. Encourage students to use images to facilitate their understanding of concepts, problems, and ideas.

5. Monitor and respond to students' use of various images.

6. Use tasks and texts that warrant the use of visual images.

By supporting students' use of images, teachers provide opportunities for students to reconsider information through mental models. Following these simple six guidelines helps students develop knowledge about how to use images as a means of representing text and making it meaningful.

Wellington and Osborne (2001) in discussing the role of communication in science point out that teaching involves a range of modes that include spoken and written words; visual representation; images, diagrams, tables, charts, models, and graphs; movement and animation of physical models; practical work; and mathematical symbols. Both science and mathematics are multi-semiotic in that these disciplines depend on more than words to communicate important ideas, concepts, and information. An investigation incorporating visualization along with problem-based methods in mathematics and science classes found that students made connections between mathematics and science data, connections between math and science language, and connections between math/science and daily life (Harnisch et al., 2005). Mental imagery enhances deep level cognitive connections that facilitate engagement, recall, comprehension, and problem solving (Douville & Pugalee, 2003). The process of forming images or visualizing can be effective in supporting comprehension of both verbal and non-verbal concepts including text with multiple abstractions, specialized symbolism, technical vocabulary, and the use of multiple-leveled visual elements.

Reading-Writing-Speaking Link

Interaction among peers reinforces memory acquisition (Winstead, 2004). In these exchanges students are complementing the information received from the teacher through discussions of concepts and ideas by their peers. This interactive discussion promotes transfer into long-term memory since these interactions allow for repeat learning of subject matter.

Writing provides unique and powerful opportunities for students to activate prior knowledge. Research has demonstrated that a curriculum that emphasizes reading and writing positively impacts the content learning of students and that students transfer this knowledge and strategies across grades and subjects (Aulls, 2003). In addition, knowledge of one process (reading or writing) reinforces the other (Irvin, 1997). Spoken language also is important in shaping students' learning. Rich language environments that promote reading, writing, and speaking help students develop deep understanding of content while supporting their development of literacy skills (Pugalee, 2005).

Teachers should consider how classroom activities, including instruction on literacy strategies, support multiple forms of communication. Instructional activities that promote reading, writing, and speaking will provide rich learning environments for students as they develop skills, processes, and understanding. These communicative processes are key components of students' actions when working in groups. Elaboration becomes a key feature to support students' work with text—both reading and writing—and with their oral communication as they exchange and construct ideas and concepts.

Some consideration of how collaborative groupings support students' construction of knowledge is worthwhile. Implementing collaborative learning should include consideration of the following key elements:

- Make the learning target explicit.
- Communicate expectations regarding collaborative behaviors as well as expectations for any products that will be produced.
- Select teams or groups. You might include specific duties for team members such as summarizer, recorder, monitor, and so on.
- Facilitate group processing. Be sure to include explicit attention to when and how to use any textual materials that are part of the activity. If necessary strategies may be taught or reviewed in groups or with the entire class before beginning the collaborative activity.
- Assess groups and individuals.

Putting It All Together

The following reading practices are adapted from those identified by teachers as valuable to student learning (Gee & Rakow, 1990). These practices are readily applicable to mathematics and science instruction. These practices can provide an ongoing base for teachers as they plan for content area reading instruction.

1. Conduct a discussion at the beginning of the year about how to use the textbook (table of contents, index, chapter headings, glossary, summaries, study aids, etc.).
2. Use questions or tasks that require students to read for different purposes: retrieve information, interpret information, and reflect and evaluate.
3. Ask students to retell or discuss what they have read.
4. Preview key vocabulary before reading.

5. Review key vocabulary after reading.

6. Provide students with guidance to a reading assignment (pay particular attention to ___ section, study the chart or example on page __).

7. Discuss the significance of graphics and examples after reading.

8. Use advance organizers and other aids to help students understand the meaning and significance of the text.

9. Explain and reinforce the use of a reading strategy.

10. Summarize information (teacher models and reinforces this practice).

It is important for teachers to reflect on successful practices so that a working repertoire of possible strategies and methods are available to assist learners in understanding text.

The following model summarizes the key steps to consider when teaching literacy strategies within the context of mathematics and science content.

Figure 3.1. Model for Teaching Literacy Skills

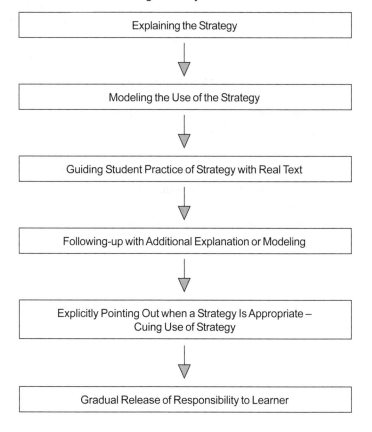

Developing literacy skills is multifaceted and requires thinking about how various literacy goals are promoted when providing explicit instruction. The considerations are important because the thinking of teachers when implementing instructional practices is never linear. The steps provide a guide to emphasize key actions, but the power lies in the movement between those steps and consideration of the important components that support literacy development.

Summary

This chapter provides some background information to guide literacy instruction in mathematics and science classrooms. This is a complex process that involves careful thinking and planning on the part of the teachers. I recognize some key concerns including the lack of significant training or professional development in reading or literacy, the number of topics in the curriculum leaves little time to include reading instruction within math and science lessons—too much to cover in one year—and then there is the pressure to produce on standardized tests. The reality is that helping students with reading will have a positive impact on student performance and literacy development and can be done within the context of teaching vital mathematics and science content. Chi (2000) points out that scientific concepts are difficult for students to understand because they are complex involving many subcomponents, abstract, and dynamic. This chapter has emphasized some key ideas about how to teach literacy while respecting the importance of the mathematics and science content. Key factors that affect students' literacy abilities were described. Addressing literacy skills helps make the content more accessible to students. The result is a dynamic interplay between literacy and content that results in engaging and dynamic instruction which helps students become more capable of learning and succeeding.

Chapter
Four

Effective Strategies
for Reading in
Mathematics and Science

Overview of the Strategies Chapter

This chapter contains strategies for reading in mathematics and science. The strategies are tools that provide procedures and methods that can be adapted in light of the reader's abilities and the nature of the reading task. The strategies are organized alphabetically. The development of each strategy includes several key sections: Reading Purposes, Description, Implementing the Strategy, and Modification and Other Considerations. Many programs organize reading strategies around whether the strategy is appropriate before reading, during reading, or after reading. The strategies in this book are identified by the purpose of reading: retrieving information, interpreting, and reflecting and evaluating. These purposes of reading adapted from the PISA assessment framework (Organization for Economic Co-operation and Development, 2003) provide a useful framework as teachers consider which strategies would be appropriate for a certain task or activity.

- *Retrieving information* involves the ability to locate information in texts. This information is usually found explicitly in the text. Retrieving information involves identifying information that is relevant to the reader's goal, looking for a specific idea, searching for definitions, identifying key concepts, and so on.

- *Interpreting information* involves the ability to construct meaning and draw inferences from written information. This purpose involves the reader in developing a general understanding of ideas or concepts in the text. Interpreting information involves identifying key components of text including the main idea and purpose, and the organization of ideas in the text to develop an understanding of the major concepts and/or ideas.

- *Reflecting and evaluation* involves the ability to relate text to other knowledge, ideas, and experiences. This purpose engages the reader in connecting information within the text to outside sources of information as well as experiences. Reflecting and evaluation involves the use of supporting evidence, concepts, and ideas from the text in various levels of reasoning including drawing information from the text, text features and language, and the evaluation of the quality and appropriateness of ideas, outcomes, and inferences.

Table 4.1 lists verbs that describe the actions inherent in these three purposes. These verbs will be helpful in identifying the appropriate strategy for a particular task.

Table 4.1. Actions in Reading Purposes

Reading Purposes	Actions
Retrieving Information	Classifying, Categorizing, Charting, Collecting, Drawing, Estimating, Graphing, Listing, Identifying cause/effect, Identifying components and the relationship between components, Identifying main ideas, Identifying patterns, Manipulating materials, Matching, Measuring, Observing, Plotting data, Prioritizing, Recording, Researching, Retrieving facts and ideas, Sequencing, Simulating
Interpreting	Comparing and contrasting, Concluding, Defining problems, Describing relationships, Drawing conclusions, Generalizing, Identifying applications, Identifying arguments and supports, Identifying cause/effect, Inferring, Making models, Reasoning, Relating, Summarizing
Reflecting and Evaluating	Applying, Assessing, Creating, Critiquing, Decision making, Describing, Designing, Developing and implementing investigations, Establishing criteria, Evaluating, Experimenting, Explaining, Hypothesizing, Identifying bias, Illustrating, Persuading, Planning, Predicting, Problem solving, Proposing solutions, Relating, Restructuring, Synthesizing, Testing

Table 4.6 lists each strategy and the reading purposes for which it is best intended.

Anticipation Guide

Reading Purposes:　Retrieving Information
　　　　　　　　　　　Interpreting
　　　　　　　　　　　Reflecting and Evaluating

Description

Anticipation guides were developed to use prediction in order to support students' comprehension of a reading passage (Head & Readence, 1986). Students activate prior knowledge as they respond to statements that reflect key concepts that are covered in the reading. For each statement, students make a choice before they read the passage and after they read the passage. Choices for the statements include strongly disagree/strongly agree; disagree/agree; likely/unlikely; and true/false. As students read, they become aware of whether the information in the text supports their original response about the concept or idea. Students should be able to defend and explain why they stuck with their original choice or why they changed it.

Implementing the Strategy

1.　Identify the major concepts or ideas that students will encounter in their reading of the selected text.

2.　Write statements that summarize these key concepts. Statements that both challenge and reinforce students' current understanding should be considered (think about what students might already know about the topic).

3.　Students either agree or disagree with each of the statements before they begin the reading passage.

4.　During reading, students consider information that supports and/or questions their initial choice for each of the statements.

5.　After reading, students once again evaluate each statement and indicate agree or disagree based on their reading. Students should be able to locate the text sections that support their decisions.

Figure 4.1. Completed Anticipation Guide on Averages.

```
                        ANTICIPATION GUIDE

Directions: Read each statement. If you believe the statement is true, put a
            check in the agree column. If you believe it is false, check the disagree
            column. Turn to your partner and share your beliefs. Then, read p. 140 in
            the textbook and complete the post reading section.


        Pre-Reading                                    Post-Reading

   Agree   Disagree          Statement              Agree   Disagree
  ____      X      1) The mean of a set of numbers is  ____     X
                      the same as the sum.

   X      ____     2) To find the mean, you first must add  X    ____
                      all the addends.

   X      ____     3) The number of addends is important  X    ____
                      in order to find the mean.

  ____     X      4) To find the mean, subtract the smaller  ____   X
                      number from the larger number.

   X      ____     5) The mean can be less than the smallest  ____   X
                      addend.
```

Modifications and Other Considerations

Teachers' modifications of anticipation guides provide a considerable opportunity for students to further extend thinking that involves reflecting and evaluating. One modification requires students to write a justification or response for why they changed their response on relevant items. A secondary biology teacher who frequently uses anticipation guides reported that he limits the number of items to five but requires a short written justification for each item. In this justification students must use information from the reading to show how their original response was correct or to show what information led them to change their choice. A fourth grade math teacher shared that she turns the statement that gave the students the most difficulty into a writing prompt for students to respond to at the end of the lesson or as part of a homework assignment. This helps to reinforce important concepts and helps students overcome misconceptions they may have formulated about the particular topic. Another elementary teacher shared that he saves the guides on his computer so that he can mix them up at the end of a unit to be used as part of a review of key concepts and ideas.

While it removes the primary influence of prediction, some teachers have students construct their own anticipation guide questions. This requires students to be familiar with the strategy. Students write responses as part of, during, or after reading activity. Teachers might also give students a theme and see what kind of statement they construct. This helps the teacher see what students already know about a topic. The teacher may then select several statements that reflect key ideas and concepts that are found in the reading.

Brainstorming

Reading Purposes: Retrieving Information

Description

Brainstorming helps students activate their prior knowledge about a particular topic. It sets the stage for the reading that will follow by helping students think about what they may already know about a topic or idea. Students generate words or phrases that are associated with a topic. These provide a means to help students anticipate key ideas or concepts that they will encounter in the text. While there are no right or wrong words, students must understand that their choices must be relevant to the word or phrase. This activity is generally done by first allowing students to generate individual words and then involving the class in sharing and recording their words. Brainstorming products can be organized in lists, flowcharts, or other graphic organizers. Brainstorming is frequently used in business settings to generate ideas and solutions to problems and issues.

Implementing the Strategy

1. Write a word or phrase on the board or overhead that is central to the reading or lesson. Keep the topic broad enough to provoke student thinking but focused on the major concept or ideas of the text passage.

2. Have students individually write down several words that come to mind that relate to the word or phrase.

3. Students share the words that they have written on their paper.

4. The teacher asks for common themes or ideas among the words (or the teacher leads a discussion of how particular words are related). Frequency of words is also noted.

Figure 4.2. Final Brainstorming Work Sample on Motion.

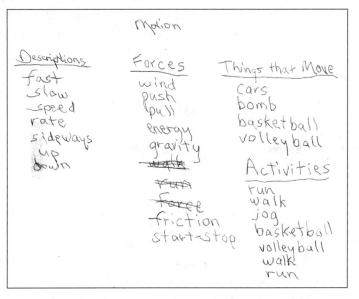

Modifications and Other Considerations

In the example above, students had brainstormed words related to motion. The teacher wrote the words on the board while the students contributed. Students in pairs then grouped words into categories. Through whole class discussion, the student pairs developed labels and modified their products. This gave students an opportunity to develop a sense of the major concepts and categories that would inform them in their reading. After this brainstorming exercise, the teacher had students read several pages from a science tradebook *Objects in Motion: Principles of Classical Mechanics* (Fleisher, 2002). Pages related to Newton's three laws of motion were copied on a transparency for the class to read. Students then completed the NASA CONNECT™ activity *Rocket to the Stars* (complete teacher guides are available at http://connect.larc.nasa.gov/). This example demonstrates how reading strategies can play a vital role in structuring lessons that draw from non-textbook print and media sources.

One modification is to have students work in small groups and go through the steps for implementing the activity. Each group can report to the class the outcome of their efforts. These group reports can serve as the basis for a discussion that introduces students to the reading. A middle grades teacher records the brainstorming results on a transparency sheet and brings it back after students have read the material and the lesson has been developed. The initial product is used to generate new words to add to the collection (which are recorded in a different color) and to engage the students in talking about how their thinking changed as a result of the reading/lesson.

Another modification is for the teacher to write three or four key words or phrases that students will encounter in their reading. This helps to focus ideas and may be helpful if the topic is one for which the teacher feels students' knowledge will be limited. The steps are similar but students respond to each of the words or phrases.

Brainstorming can be extended by having students create graphic organizers or maps with their word choices. This helps students see relationships among words and ideas. Students can use the graphic organizer to add information as they read the text. This level of modification extends students' thinking and engages them in reflecting and making decisions about commonalities and associations between concepts and ideas encountered in the reading (see semantic mapping).

Cause & Effect

Reading Purposes: Retrieving Information
                   Interpreting
                   Reflecting and Evaluating

Description

Students need to develop critical thinking skills that help them identify cause and effect relationships. Many of these relationships will be stated explicitly in text while others require students to think about the relationship between events. Cause and effect relationships describe why something happens. The **cause** is the event that triggers something to occur or "the why things happen" and the **effect** is the result or "the what happened." Cause and effect relationships identify potential results of an event, problem, or issue. This strategy helps students recognize the text structures that are used to express this type of relationship. Students should be familiar with common words that signal cause/effect relationships such as because, so, consequently, thus, since, as a result, therefore, due to, nevertheless, so that, if, then, and so on.

Implementing the Strategy

1. Allow students to read or explore the text with one of the purposes of reading to identify a cause and effect relationship.

2. Students complete the cause and effect organizer either individually, working in small groups, or as a whole class.

3. The structure of cause and effect relationships should be reinforced by having students state the relationships verbally and/or in writing. If… then… expressions will help students understand the structure of these relationships. As students become comfortable identifying and working with such relationships, they may use other ways of expressing the relationships.

Figure 4.3. Cause and Effect Graphic for Three States of Water

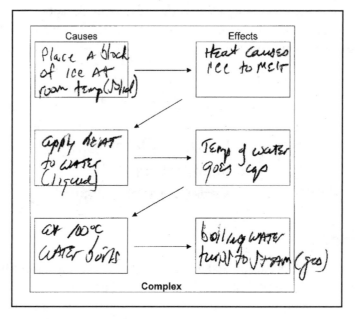

Modifications and Other Considerations

This example shows a complex cause and effect relationship. The organizer explores the cause and effect relationship, "What would happen if ice (solid state of water) was left at room temperature and then heated to boiling?" The strategy can also be used for simple relationships by using the first two blocks of the grid. Additional blocks can be added if the cause and effect relationship involves additional steps or stages. These multi-stage events are sometimes referred to as "causal chains" where one event leads to another. Sometimes an event (cause) has several effects. For example, deforestation is an event that has several effects such as erosion, flooding, changes in climate, and reduction in diversity of wildlife.

One modification of the strategy is to cut out the blocks with the information about the cause and effect relationship and have students match the correct pieces of information. This can be especially effective if students have several cause and effect relationships which they have studied. Teachers might also fill in parts of the organizer and allow students to complete the missing information. One teacher puts all the appropriate statements on an overhead and has students complete the organizer after they have completed the reading of the material. Students seeing the statements before reading helps focus their attention to important relationships by focusing their awareness on the potential cause and effect relationship(s) expressed in the reading.

Cause and effect relationships are very important in science, but such relationships are also common in mathematics. A simple example involves multiplying the sides of a figure (such as a rectangle) by the same number, such as 2. The result is a figure that is similar to the original (that is the angles are congruent to the original rectangle and the sides are proportional).

Cloze Vocabulary Exercise

Reading Purposes: Retrieving Information

Description

Cloze procedures were originally used as a test of reading comprehension formed by creating fill-in-the-blank activities in which words were removed from the text. This activity is greatly modified but still promotes students' use of context clues to select appropriate vocabulary words. The activity helps students predict the word that should appear in the blank. The activities are designed specifically to reinforce important words and concepts. Such activities can be effective in supporting students who have difficulty with key vocabulary such as students who are English language learners (Pugalee, Brailsford, & Perez, 2005). Unlike cloze procedures used to assess reading readiness, this modification is designed to assist students in recognizing key vocabulary words from a selected reading passage. As students differentiate between key words in making a selection, they build comprehension skills by using context clues to help determine the best word choice.

Implementing the Strategy

1. Select a piece of text that students will read or work with as part of a lesson.

2. Identify key vocabulary words that appear in the text.

3. Design fill-in-the-blank sentences that use the key vocabulary word. The sentences should maintain the contextual information from the lesson.

4. Create a list of possible words to be used in the blanks. Add several words that are related to the lesson or topic but are not to be used in the exercises.

5. Students complete the exercises either as they read the text selection or as a post reading activity.

Figure 4.4. Cloze Vocabulary Exercise.

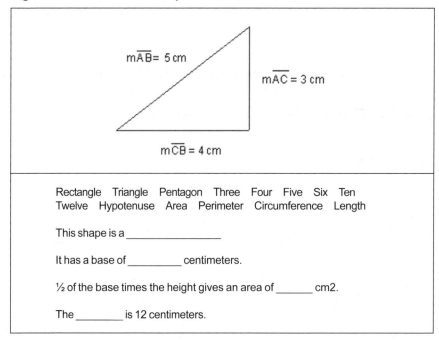

Rectangle   Triangle   Pentagon   Three   Four   Five   Six   Ten
Twelve   Hypotenuse   Area   Perimeter   Circumference   Length

This shape is a _____

It has a base of _____ centimeters.

½ of the base times the height gives an area of _____ cm2.

The _____ is 12 centimeters.

Modifications and Other Considerations

This strategy can be used with work in small groups or pairs. Notice in the example that the vocabulary term "base" is given in the sentence but the students must select the correct number for the length. This type of activity still requires students to understand the vocabulary being used but places it in somewhat of a different context. Teachers who are working with students who have less difficulty with word recognition and vocabulary use may modify the activity by not providing a list of possible words to fill in the blanks. More complex sentences may be written including sentences where more than one key word is omitted. The complexity of the sentences depends on the level of the students for which the activity was designed. Such exercises when developed for mathematics and science concepts may often include a diagram or illustration. Students may also benefit by developing their own sentences as a post reading activity done after a lesson.

Compare & Contrast Summary

Reading Purposes: Retrieving Information
                               Interpreting
                               Reflecting and Evaluating

Description

The compare and contrast summary method is designed to provide a way for students to organize information that compares and contrasts two concepts. The strategy reinforces students' work with text structures that use comparisons and contrasts to develop key concepts. Students' thinking that involves comparing and contrasting involves actively working with two topics or concepts. The strategy supports students in asking key questions about the topics under consideration. What are the characteristics or qualities on which the topics are being compared? How are they alike? How are they different? The strategy is extended by providing a Summary Writing Frame to guide students in putting key ideas in written form (Ambruster & Anderson, 1989). The writing framework is especially effective in supporting students in compare/contrast writing which is one of the more difficulty types of expository writing.

Implementing the Strategy

1. Identify several key concepts from the text, labeling the columns in the grid with these concepts or ideas.

2. Identify the features that students will use to develop their comparison/contrast of the concepts or topics. One feature or characteristics is placed on each row on the left side of the grid.

3. Students read the text and summarize key information about the topics. The grids provide a place for students to record ideas, facts, and concepts that show similarities and differences relative to the features or characteristics under consideration.

4. There needs to be some type of reporting or sharing so that the accuracy of ideas and information is checked.

5. Students use the information to complete succinct statements comparing and contrasting the ideas. The Summary Writing Frame provides a tool to guide students in their writing. Teachers find that the writing stage is most effective after some discussion or checking for understanding (step 4).

Figure 4.5. Compare and Contrast Summary.

	Planets	**Moons**
Revolution	Bodies that revolve around a star such as our Sun and reflects light but does not produce energy through nuclear fusion. Comets and meteoroids are not planets.	Revolve around a planet and are called satellites. Satellites that revolve around planets are referred to as moons, in reference to Earth's satellite which is called the Moon.
Size	Smaller than the stars that they orbit. Planets range in size from less than half the size of Earth to giants that are thousands of times larger.	Smaller than the planet they orbit. Earth and Pluto have the largest moons in relation to the planet's size. The moon of Jupiter is larger than the planet Mercury.

Summary Writing Frame:

Example Comparison/Contrast Summary Writing Frame:
_____ and _____ are alike is several ways. First, _____. Second, _____.
Finally, both ___ and _____ _____. Yet, _____ and _____ are
different. First, _____, while _____. Secondly, _____ while _____.
Finally, ____ _____, while _____.

Figure 4.5. Compare and Contrast Summary. *(Continued)*

Planets and moons are alike in several ways. First, they both revolve around other heavenly bodies. Second, planets and moons are smaller than the bodies around which they revolve. Yet, planets and moons are different. First, planets revolve around a star while moons are satellites that revolve around planets. Secondly, planets and moons vary greatly in size with some planets thousands of times larger than Earth. While moons are smaller than the planets they orbit, a moon may be larger than some other planet.

Modifications and Other Considerations

This strategy can also be used as either an individual, group, or whole-class activity. The strategy works well for readings that are assigned for outside class. Columns and rows can be added for readings that involve more than two concepts and/or features. For example, a compare/contrast summary of mathematics properties: associative, commutative, and distributive would need three columns and four rows (addition, subtraction, multiplication, division). The Summary Writing Frame provides a general structure. Notice how the format changes somewhat with the writing summary comparing and contrasting planets and moons along the features of revolution and size. Teachers may decide to modify the frame if the wording will become awkward. Students will be able to modify the general frame once they are more familiar with compare/contrast writing. The frame provided has room for three features. The reading may only involve two or sometimes more than three.

Concept Circle

Reading Purposes: Retrieving Information
Interpreting
Reflecting and Evaluating

Description

This strategy assists students in understanding conceptual connections between words or phrases. Concept circles support metacognitive behaviors as students are required to reflect on the relationships between the words, categorize concepts and ideas, and construct descriptions of complex relationships and information.

Concept circles encourage active learning as students are prompted to take implicit and fuzzy associations and make them into explicit and precise linkages. This promotes the type of knowledge that is essential for recall and comprehension (Fisher, Wandersee, & Moody, 2000). Concept circles are also effective in helping students compare attributes and examples that are related to important mathematical and scientific concepts and processes. Circles can be divided into any number of sections depending on the number of concepts discussed in the text.

Implementing the Strategy

1. Identify a concept that is central to the reading.

2. Divide a circle into equal parts depending on the number of attributes or words that are needed to extend understanding for the concept.

3. As students read, they identify key attributes, characteristics, properties, relationships, or other information that describes that particular concept. One word or a short phrase is used to summarize each key piece of information that is related to the concept.

4. Follow-up discussion or instruction allows for the modification of the concept circles so that they are accurate and complete.

Figure 4.6. Concept Circle for Three-Dimensional Figures.

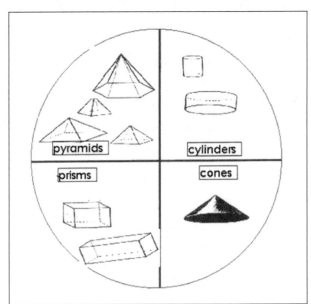

Modifications and Other Considerations

In the example, note that the student had to place the appropriate three-dimensional figures in the appropriate section of the circle. Most concept circles use words or phrases to record ideas about important concepts. This example demonstrates the important role of diagrams and illustrations in mathematics. They are equally important in studying science. The student had to be able to identify the characteristics of the figures based on the reading in order to place them correctly in the sector for that particular shape.

A popular modification is for the teacher to include one word, phrase, diagram, or illustration that is not related to the others. The students then identify the unrelated word and describe how the other words are related and a rationale for why they feel their selected word does not fit with the others. Such modifications promote a connection between reading and writing allowing students to draw on other parts of the text, other readings, and prior knowledge to defend their reasoning. Other modifications include providing the attributes or relationships and have students identify the concept. Some teachers provide circles with two or more sections completed and ask students to fill in the remaining sections. This requires students to identify the relationship and to decide what important information has not been included. Concept circles with the words or phrases in the sections can be used effectively as part of a writing assignment or an open-ended assessment where students must describe how the words/phrases are related.

Concept circles can also be used in collaborative learning settings. In these collaborative learning environments, students are encouraged to discuss, describe, and consolidate thinking about the meaning of concepts and the relations between them (Fisher, Wandersee, & Moody, 2000). Students should have some opportunity to revisit their products at the end of the collaborative experience. Sharing the concept circles forces students to synthesize and compare ways that knowledge is represented and helps students use common language so that ideas are communicated clearly and precisely. It is important that students' concept circles reflect accurate and complete information that is central to the reading.

Concept Synthesis

Reading Purposes: Retrieving Information
Interpreting
Reflecting and Evaluating

Description

The concept synthesis strategy supports students in differentiating between ideas presented in text as they identify main ideas and concepts. Students obtain ownership of those concepts by writing them in their own words. The final component of the strategy reinforces the reading-writing link by engaging students in writing a short synthesis of the concept based on the information they have developed that explains why those concepts or ideas are important. This strategy is especially effective with dense texts, but requires careful modeling with students so that they can be successful in using it on their own. Modeling the strategy with the whole class is important. Students can underline or highlight key ideas as the teacher models how the strategy is used.

Implementing the Strategy

1. Identify several key concepts from text, providing page numbers or locations of the information. State these concepts as they are presented in the text.

2. Write the concepts identified in the first column in your own words. Be careful to use appropriate vocabulary so that key information is preserved.

3. Think about how your concepts are related and write several sentences in the last column that relates those concepts and explains why they are important.

4. As a class, pair, or group activity compare products assimilating information and concepts into a final class or group product.

Figure 4.7. Concept Synthesis Used in Middle Grades Lesson on Cells.

Major Concepts Synthesis

Identify the most important concepts from your reading. Think about how you would explain those concepts to someone who had not read the text. Record your information in the table.

Key Concept (Brief Statement with Page Number)	Concept in Your Own Words	Three or Four Sentences Relating the Concepts (Explaining Why They Are Important)
ALL living things are made of cells. (1)	Some living things are one cell and others have trillions of cells.	Every living thing is made of cells. Cells are very specialized.
Some people think of cells as a balloon filled with fluid. (2)	Un like a balloon cells let things pass in and out.	The plasma membrane allows water & food to pass in & out.
Cells are amazing—they do many different things.	Cells are programmed to do a lot: carry oxygen, make new cells, filter/transport.	Some cells carry oxygen. Some carry signals to our brain. Cells are very busy.

Modifications and Other Considerations

This strategy can also be used as part of a whole-class activity. Teachers find that providing students with the key concepts already identified in column one is a good way to help students become familiar with the strategy and to support them if the text is "dense" or contains a lot of distracting ideas or information. Students must still process the text and understand the key ideas to restate them in their own words and ultimately write a short synthesis relating the concepts and ideas. Teachers also report doing the first two columns as a class activity and having students complete the synthesis part individually, sometimes as a homework or out-of-class assignment. Mathematics teachers find that concepts and ideas often involve symbols and/or diagrams. Students need explicit instruction so that they describe how these representations are related to the main ideas or concepts in the text. As students engage in this strategy they monitor the development of meaning through identifying key concepts and themes in the text, and determining how those ideas fit together through a short writing experience.

Cubing

Reading Purposes: Retrieving Information
                       Interpreting
                       Reflecting and Evaluating

Description

Cubing assists students in comprehending key topics or concepts from multiple perspectives. The cube provides a guide for students to consider mathematical and scientific concepts and ideas from different vantage points and reinforces students' critical analysis of text. Each side of the cube has a different perspective written on it. Students explore a topic from each of the six perspectives. Teachers might write stem questions or ideas on the cube for each perspective, or post the ideas on chart paper or the overhead. Students will need the descriptions of each perspective in order to move successfully through the six perspectives.

- **Describe it.** Students use the text or other informational sources to describe the topic or concept. The description may involve the use of definitions, theorems, characteristics, or properties.

- **Compare it.** Students compare key concepts or ideas that are part of the lesson or to concepts or topics they have learned previously. The goal is for students to focus on similarities between concepts and ideas.

- **Associate it.** Students relate the topic to other issues, concepts, ideas, events. The goal is to help students connect the topic to other ideas and concepts.

- **Analyze it.** Students focus on key components or details that are central to the concept or topic.

- **Apply it**. Students focus on applying or using the concept or idea.

- **Argue for or against it**. Students reason about some key fact. The reasoning may focus on how the concept or idea works.

Implementing the Strategy

1. Students working in small groups will toss the cube. (The diagram below is a net of a cube. It can be folded and taped to form a cube). Students should read the text information prior to getting into groups.

2. Students explore the concept using the perspective that comes up on top of the tossed cube. It is important for the group to record their ideas. Teachers often limit the time for groups to respond to each of the perspectives. The teacher may tell groups when to wrap up their discussion and when to toss the cube again.

3. The cube is tossed again. The new perspective is discussed. If a perspective that has already been discussed comes up, the cube is tossed again.

4. The process is repeated until all six different perspectives are discussed in the groups.

5. Whole-class discussion combines key ideas from each of the six perspectives. This reporting out process is important in extending students ideas and making sure that all the key information from the lesson is discussed and developed.

Figure 4.8. Cubing Reading Perspectives.

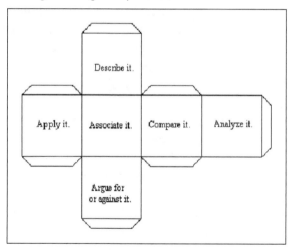

In a lesson on percentages, students generated the following ideas based on their work with the text in small groups:

Describe it: A percentage is how many parts out of 100.

Apply it: If I have ¼ of something that is the same as 25/100 which is 25%.

Associate it: One way of thinking about percents is to think of a grid with 100 squares. What part is shaded to show the number of parts? In

¼, I would shade 1 out of every 4. If this is done with 100 squares, I will have shaded 25 of the 100 or 25%.

Compare it: Percents, fractions, and decimals can represent the same number. ¼ is .25 is 25%.

Analyze it: The best way to think of percents is as a fraction with 100 as the denominator. Percent means "parts per 100."

Argue for or against it: Percents are important because they give a standard base (100) for comparing things. We use them all the time: sports, shopping, business, and even in school when we talk about grades.

Modifications and Other Considerations

The teacher may toss the cube in front of the class to promote whole-class discussion. The teacher may also toss the cube (or get a student to toss it) and call out the perspective in order to guide small group discussion of the perspective. Teachers sometimes toss the cube and complete one or two perspectives with the entire class before the groups explore the remaining perspectives. Teachers may use the cube as a post-reading strategy to get students to think about the information they have read. Cubing can also be used independently by students to help them explore text and understand it from important perspectives. Teachers might also modify the six perspectives to fit a particular lesson.

Cubing can also be used as a writing strategy to stimulate students' thinking and synthesis about a topic. It can be used effectively to integrate writing into mathematics and science (Pugalee, DiBiase, & Wood, 1999). For example in problem solving the six perspectives could be organize, describe, analyze, predict, summarize, and explore. The cube becomes a guide to help students make sure they can communicate their thinking. Students could write summaries or paragraphs related to a particular perspective and then work in groups to consolidate their ideas and writing into a larger written product.

Cue Cards

Reading Purposes: Retrieving Information

Description

This strategy is frequently used in game format. It is designed to help students become familiar with important terms and concepts. The strategy serves as a good review or quick assessment of student learning. Students are required to think about information or concepts and identify the correct label, word, or illustration that best identifies that concept or idea. As such, students are comparing and contrasting basic ideas to make their choices. Teachers can modify the strategy to students' levels of knowledge. The cards used in the strategy can be replaced with sets designed to meet the instructional goals of a lesson or unit.

Implementing the Strategy

1. Construct six to eight index cards with an important concept, problem, idea, or other important information from a lesson or unit.

2. Prepare cue cards that match the information on the cards created in step 1. You may also add cue cards that have no matches or have more than one possible cue card that matches.

3. Give students the set of six to eight cards created in step 1.

4. Show and read the cue cards aloud to the students. As each is presented, students will hold up the matching response.

5. Follow-up with discussion or questioning related to concepts as misconceptions or gaps in knowledge are revealed.

Figure 4.9. Example Matching Cue Cards from Circle Introduction.

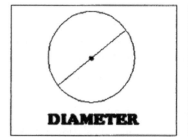

DIAMETER

A line that connects 2 points of a circle by going through the center of the circle.

The longest cord of a circle.

Modifications and Other Considerations

Additional sets of cards can be constructed so that students can work in small groups or pairs. Some teachers add cards as concepts are covered in class. Teachers may decide to use models, diagrams, or illustrations instead of words if those types of representations are important in helping students understanding key course concepts. For example, a science teacher working with the parts of a plant might select to have illustrations of plant parts on the cards with the matching card providing a fact or detail about the particular part.

The activity may also become a student-generated activity. Teachers may want to provide the concepts, ideas, or terms for which students might create matching cards. Students can create card sets to be used throughout a lesson for review of concepts. One science teacher reported that she had students construct a set of cards for each unit. Each student was given two index cards when a new concept was covered. Students created their cards under the direction of the teacher and added them to the appropriate deck (one containing concepts or the one containing the matching term) which was kept in a sealed plastic bag. This allowed for each student to have a set of cards to use with peers or when the teacher did an entire class cue card activity. The process of constructing the cards was an important learning strategy in that it provided a summary of some key concepts covered in class. Having the students construct sets of cards was also more time efficient.

Fact Summary

Reading Purposes: Retrieving Information
Interpreting

Description

Mathematics and science textbooks contain information text. Organizing all of these facts so that the information is helpful in approaching learning tasks can be daunting for students particularly when they need to extract lots of information for several pages or chapters. Though these facts and details are usually presented in a straightforward manner, they are conceptually dense. Technical terminology, charts, graphs, examples, and formulas often contain facts that students need to apply or to understand major concepts. Without a strategy, students may not be successful summarizing the important factual information presented in the text. This strategy makes reading purposeful and active as students record information as they encounter it in their reading. The Fact Summary strategy helps students organize ideas and information in one place so that it becomes a reference guide. The strategy helps students identify ways of organizing several facts and assists them in identifying what information is important.

Implementing the Strategy

1. Introduce the strategy at the beginning of the lesson or lessons in which students will need to organize information they encounter.

2. Review the components that will help focus students' attention on key ideas. These components are the labels across the columns and rows.

3. Review information periodically as it is covered in the lesson or appears in the text.

4. Once the table is complete, review the information with students for accuracy.

5. The Fact Summary table can be used as a reference to help students in applying the information as it is being completed as well as when it is finished.

Figure 4.10. Fact Summary for Regular Polygons.

Regular Polygon Name	# of sides& angles	# of triangles	Total degrees	Degree measure of each interior angle	Number of Diagonals
triangle	3	1	180	60	0
quadrilateral	4	2	360	90	2
pentagon	5	3	540	108	5
hexagon	6	4	720	120	9
septagon/heptagon	7	5	900	128.5	14
octagon	8	6	1080	135	20
nonagon	9	7	1260	140	27
decagon	10	8	1440	144	35
n-gon-formula	n	n-2	$(n-2)(180)$	$\frac{(n-2)180}{n}$	$\frac{n(n-3)}{2}$

n = number of sides

All polygons - sum of exterior angles equals 360°

Modifications and Other Considerations

This strategy is intended to help students organize information and facts. The strategy can also be used as a tool for recording conclusions made from explorations and experiments. Teachers may elect to have a class Fact Summary grid on the board or posted in the classroom that can be completed as relevant information is encountered in the lessons. The finished product provides a ready reference for later work and as a handy guide for students as they review material to prepare for assessments. Teachers can add columns that include examples, illustrations, or definitions. One teacher reported using a partially completed table as an assessment which required students to fill in the missing information.

Five-Minute Pause

Reading Purposes: Retrieving Information
                                 Interpreting
                                 Reflecting and Evaluating

Description

The five-minute pause provides a block of uninterrupted time for students to recall as much detail related to key ideas and information from the text as possible. The strategy is intended to be FORMATIVE in nature and not an assessment of students' reading or comprehension; however, the strategy does promote regulation of learning. Over time, the use of the strategy helps students learn how to focus on key ideas and information as they read text. By stating the basic propositions and concepts from the text, students review and reconstruct meaning from the text passage. Students may have difficulty retelling or recounting key ideas. It is important to model how pausing, reflecting, and retelling assists in recall and remembering information. Many teachers emphasize that students should be allowed to revise the products of the five-minute pause so they can incorporate anything missed or misunderstood in their initial reading. This follow-up emphasizes the ongoing development of reasoning and meaning related to the concepts and ideas in the text.

Implementing the Strategy

1. Pause students after a section of text and have them get into groups of three or four. Teachers advise against using pairs since if one student is experiencing difficulty the sharing is less robust.

2. Students spend about three minutes focusing on the key ideas from the reading. Individually, students write down ideas raised in the discussion. The point is for them to identify key points in the text.

3. Students spend about two minutes individually adding additional information or extending ideas so that they are clearer to them.

4. Repeat as necessary with other blocks of text. The size of the text blocks for each iteration can vary depending on the complexity of the concepts and the abilities of the students. Allow students to revisit their information after the lesson is complete, revising ideas and adding information emphasized in follow-up instruction and/or discussion.

Figure 4.11. Student's Summaries for Five-Minute Pause on Graphing Linear Equations.

1	An equation like $2x + y = 4$ is linear. (NO exponents with x or y)
2.	The graph of this type of equation is a line-linear.
3.	There are an infinite number of points (x, y) that make the equation true.

Modifications and Other Considerations

The five-minute pause is an extension of the more common three-minute pause stressing reflecting and telling. Many teachers found that five minutes was more productive than three if it required students to revisit written notes from the group sharing. Most teachers who use five minutes allow students to use their text to add information during the last two minutes.

Teachers who plan to use more direct instruction may focus only on the first part: read, pair, share. Regardless, the opportunity for students to pause, reflect, and recount key concepts and ideas is productive in their comprehension of the text. Teachers might also choose to engage students in whole-class discussion which might include construction of charts, diagrams, or drawings that summarize the ideas emerging from the reading and sharing. Both of these alternatives respects the intent of the strategy by allowing additional processing time for students to revisit their initial formulations, check the accuracy of their ideas, and add any information that further clarifies the concept.

Frayer Model

Reading Purposes: Retrieving Information
Interpreting
Reflecting and Evaluating

Description

While the Frayer model (Frayer, Frederick, & Klausmeier, 1969) is most often touted as a vocabulary development strategy, limiting the strategy only to vocabulary development fails to utilize the strategy to its fullest potential as a learning tool. The model requires students to define a word; describe characteristics, properties, or facts; provide an example; and provide a nonexample. The strategy helps students learn unfamiliar vocabulary and concepts. The activity requires students to think about what the concept is (examples) and what it is not (nonexamples). This level of analysis involves both interpreting and evaluating information often extending their understanding beyond the information presented in the text.

Implementing the Strategy

1. Select a key word from the text. Provide some background relative to the term before students read the selected text.

2. Students can work individually or in pairs or groups to define the concept and provide details or properties. If a concept has multiple properties or facts, teachers might give different pairs or groups different key concepts.

3. Have students share their completed products with the entire class allowing students to modify their original work to incorporate omitted or new information.

Figure 4.12. The Frayer Model Used in a Fifth-Grade Lesson on Polygons.

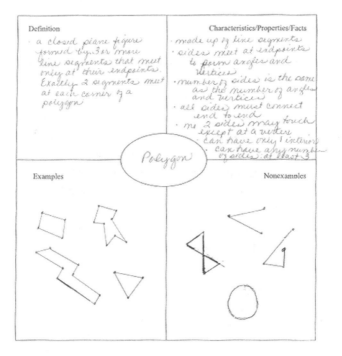

Modifications and Other Considerations

Identifying examples and nonexamples is a powerful tool for developing understanding of concepts. Initially, students will need support and modeling to think about nonexamples. Teachers might provide one example and one nonexample and have students come up with additional ones. The grid with characteristics can be modified to fit the nature of the text. For example, the block can be used to identify steps in a mathematical process such as how to find an equivalent fraction or to identify the steps in the water cycle as part of a science lesson. Another way to modify the strategy is to combine examples and nonexamples in one grid and leave the last grid for a practice problem in mathematics, application of a scientific concept, or a diagram or illustration related to the concept (see Stephens & Brown, 2005). This strategy lends itself well to the use of diagrams, illustrations, graphs, tables, and other ways of representing a concept.

Highlighting

Reading Purposes: Retrieving Information
Interpreting

Description

Students are frequently asked to highlight (or underline) while they read. The result sometimes is a mass of highlighted text with no real differentiation, almost every sentence is selected. Highlighting is a strategy that assists students in identifying main ideas or key concepts along with supporting details. The goal is to help students target and condense information, improving their reading comprehension. Highlighting provides an active reading strategy that helps focus the reader to identify key concepts or topics and their supporting ideas. The strategy limits the amount of text that students must attend to in order to understand the basic idea in a block of text. Highlighting can also be an efficient tool for reviewing text. This strategy is also highly versatile and can be modified to fit specific instructional goals.

Implementing the Strategy

1. Select a block of text that contains one or more key ideas (break large passages into smaller blocks).

2. Identify the focus of the highlighting such as main idea(s), key words and supporting ideas, definitions, or concepts and examples.

3. Highlight one or several words according to the focus. Do not highlight multiple sentences, but select key words and phrases that capture the idea.

4. Review highlighted material and summarize the information selected. It may be helpful to write a summary or to make follow-up notes in the margin of the text.

Figure 4.13. Highlighting of Text Section on Radiation.

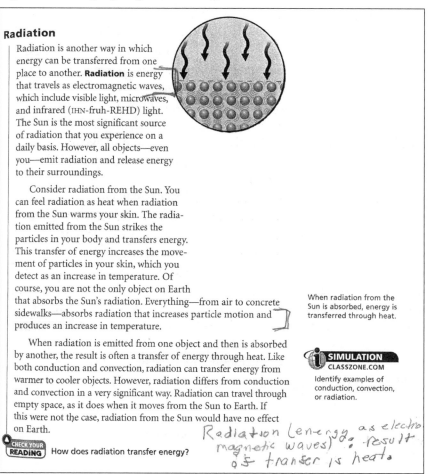

Radiation

Radiation is another way in which energy can be transferred from one place to another. **Radiation** is energy that travels as electromagnetic waves, which include visible light, microwaves, and infrared (IHN-fruh-REHD) light. The Sun is the most significant source of radiation that you experience on a daily basis. However, all objects—even you—emit radiation and release energy to their surroundings.

Consider radiation from the Sun. You can feel radiation as heat when radiation from the Sun warms your skin. The radiation emitted from the Sun strikes the particles in your body and transfers energy. This transfer of energy increases the movement of particles in your skin, which you detect as an increase in temperature. Of course, you are not the only object on Earth that absorbs the Sun's radiation. Everything—from air to concrete sidewalks—absorbs radiation that increases particle motion and produces an increase in temperature.

When radiation is emitted from one object and then is absorbed by another, the result is often a transfer of energy through heat. Like both conduction and convection, radiation can transfer energy from warmer to cooler objects. However, radiation differs from conduction and convection in a very significant way. Radiation can travel through empty space, as it does when it moves from the Sun to Earth. If this were not the case, radiation from the Sun would have no effect on Earth.

When radiation from the Sun is absorbed, energy is transferred through heat.

SIMULATION
CLASSZONE.COM
Identify examples of conduction, convection, or radiation.

CHECK YOUR READING How does radiation transfer energy?

Radiation (energy as electromagnetic waves) do. result of transfer is heat.

Modifications and Other Considerations

Notice the minimal amount of text that is highlighted in the example. Highlighting helps students by causing them to focus on a few major ideas presented in the text. In this classroom, the teacher has students use highlighting anytime they read a block of text. Students are trained to highlight sparingly paying attention to the one or two major points in the text. Students are also encouraged to write a very concise statement about the major idea. Note that the publisher of the text is this example uses boldface print to help students focus on key vocabulary. The text also emphasizes reading for understanding. In this example, students are presented with a key question to help them focus on the key idea presented in the block of text (McDougal Littell, 2005).

Several teachers pointed out that simply highlighting isn't sufficient and that highlighting should be used in conjunction with other strategies that support students in critically reviewing and understanding ideas. Highlighting works well with other strategies supporting comprehension. Some stress that highlighting should lead to writing and summarizing as part of the strategy.

One teacher has students work in cooperative groups or pairs after reading and highlighting. This gives students an opportunity to discuss what was highlighted and why as well as to develop a summary of the key ideas from the highlighted text. A less effective but valuable alternative is to do whole class highlighting which can be done by copying the text onto a transparency and highlighting with input from the class.

Students can use multiple colors to code during highlighting. Students might use one color for concepts or ideas and another color for supporting information. Teachers might identify key concepts that will be encountered in the reading and students can use different colors to highlight important information for each of the concepts or topics. One teacher recommends that definitions appearing in text have only the word highlighted and then the definition in brackets or parenthesis. Some teachers use "sticky" notes to support highlighting by having students summarize the information highlighted in a few words or by constructing an essential question that is answered by the highlighted text. Stickys are also a viable alternative when the textbook cannot be highlighted by students because it is used multiple times.

Inquiry Chart

Reading Purposes: Retrieving Information
Interpreting
Reflecting and Evaluating

Description

Inquiry charts (I-Charts) provide a framework assisting students in focusing research on a topic and organizing their thinking. As such, the I-chart also serves as a useful organizational tool for writing. Inquiry charts help students generate questions related to a topic, identify what is already known about the topic, and organize information that comes from multiple content-area sources. I-charts help make reading strategic, fostering the development of good research skills, and the synthesis of information from multiple text sources.

The organization promotes synthesis of ideas from a variety of topics assisting students in interpreting differences in how information is presented and making judgments on the usefulness of that information in addressing questions that relate to a particular topic. The strategy also helps students resolve competing ideas and generating new questions that emerge from the reading.

Implementing the Strategy

1. Identify a topic for research. The topic should be narrow enough that the major ideas can be addressed through three or four central questions.

2. Formulate three or four questions that seem important to the topic being studied.

Figure 4.14. Inquiry Chart on Prime Numbers.

Inquiry Chart: Prime numbers
(Topic)

	Guiding Questions			New Questions
	What is a prime number?	What is the smallest prime number?	What is the biggest prime number?	
What We Know Now	a number that can't be divided.	0 can't be divided	It is big	
Source 1:	Whole numbers greater than 1 whose only factors are 1 and itself	0 and 1 are prime or composite		How do you find the next one?
Source 2: encyclopedia Britannica	Positive integer greater than 1 divisible by 1 and itself	sequence begins 2,3,4,7... follows no pattern	There are infinitely many.	How do you find the next one?
Source 3: Prime Pages (internet)	No factor except itself and 1.		(234843843...) found on Dec. 15 2005	How big is the largest prime?
Source 4:			(9.1 million digits long)	
Summary	A prime number can only be divided by 1 and itself.	So the smallest prime number is 2.	The largest known prime number is very large.	

3. Write the questions at the top of each of the columns of the inquiry chart. Use the first row of the chart to record what you already know relative to each question.

4. Use multiple sources to locate information that addresses each question. List these multiple sources on the left-hand side of the inquiry chart. (Teachers should consider requiring students to keep more detailed bibliographic information on another sheet or use the back of the inquiry chart sheet.)

5. Record the information in the grids in summary form.

6. For each question, formulate a summary of the information for each question. Write this summary in the last row of the chart.

7. The information, particularly the summaries, provides a context for writing about the topic or for engaging in discussions about key concepts.

Modifications and Other Considerations

Some teachers engage the class in identifying potential questions about a topic. Students might also be engaged in brainstorming what they already know about a particular topic. This provides structure and guidance for those students who may have difficulty formulating questions or identifying an appropriate topic for consideration. Some teachers assist students in completing one row of the chart from the common classroom text then assigning individual questions to small groups of students. Multiple groups might be assigned the same question and given an opportunity to compare their responses. Students or groups with the same question may consult different sources thus adding to the richness of the information.

Inquiry charts can be used individually, in small groups, or in whole class contexts. Large inquiry charts can be drawn on poster paper for use in small groups or on the chalkboard for use with whole classes. Whole-class completion of a topic using an inquiry chart is a useful modeling tool when students are first introduced to the strategy. Students might record information responding to the specified questions from sources on sticky notes and place those on the large inquiry chart. Summaries can be generated individually from the information or can be done in smaller groups.

Jigsaw

Reading Purposes: Retrieving Information
Interpreting
Reflecting and Evaluating

Description

Jigsaw is a strategy that depends on the use of cooperative learning (Aronson et al., 1978).There are many variations of the idea but each has some basic principles (the version here is more akin to what is referred to as Jigsaw II). As the term implies, the strategy requires students to become knowledgeable on a particular topic (experts) (or a particular block of text) and then "teach" that information to others in the group who depend on the knowledge in order to develop a cohesive understanding of a concept or idea. Students are assigned to a home group but leave that group to become an expert on the topic that is assigned (or a particular block of text). For example, a large section of text might be divided into four to six blocks with an expert group responsible for reading and understanding the important information in each block. The expert group members work together to develop an understanding of their topic (or text passage). Students must make decisions on what is essential so that everyone feels confident in presenting that information to their home groups. Students move back into home groups who depend on each member to share so that they have all of the information to the concept or idea, which gives them all the pieces to the puzzle. It is recommended that for the first few times using this strategy that students be given short blocks of text or simpler topics.

Implementing the Strategy

1. Divide the class into groups with 4 to 6 members depending on the number of topics to be researched or the number of blocks of text in a selected reading. This is the Home Group.

2. Assign each student in the home group one of the topics or one of the blocks of text. Make sure that each topic or block of text is assigned to at least one member of the home group. If there are more group members than topics or text blocks, more than one member may be given the same topic.

3. Students leave their home groups and meet with students who have been assigned the same topic or block of text. These groups are Expert Groups. Students are responsible for understanding their topic or text, discussing important ideas, and deciding on what information to present as experts.
4. Students leave their expert groups and reassemble in their Home Groups. Students "teach" the information they developed in their expert group. Team members listen carefully and should be encouraged to take notes as each topic or text block is taught.
5. Students produce a product based on their understanding of the topic. This might be an activity, information for a presentation, notes on the topic for discussion, or a quiz.

Figure 4.15. A Student's Notes from Waves Lesson for Expert Group Processing.

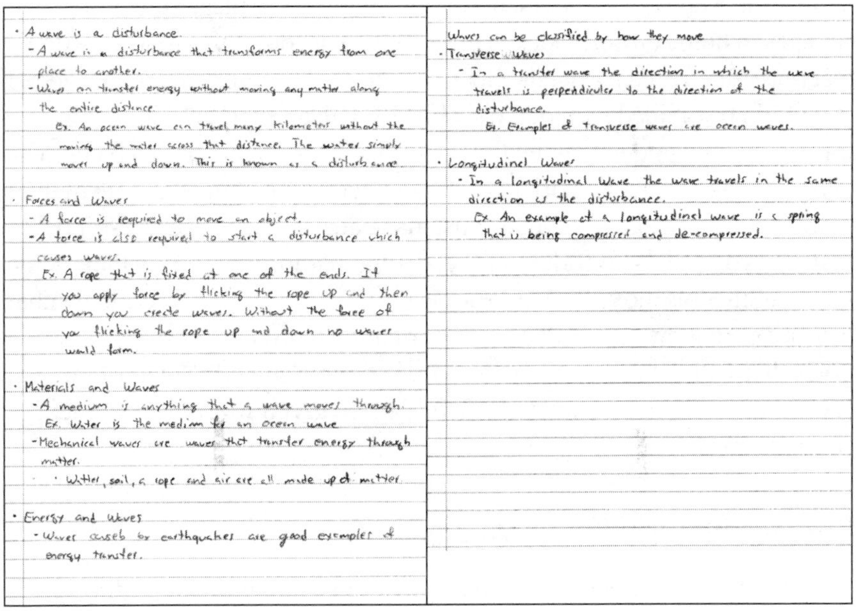

- A wave is a disturbance.
  - A wave is a disturbance that transforms energy from one place to another.
  - Waves can transfer energy without moving any matter along the entire distance.
    - Ex. An ocean wave can travel many kilometers without the moving the water across that distance. The water simply moves up and down. This is known as a disturbance.
- Forces and Waves
  - A force is required to move an object.
  - A force is also required to start a disturbance which causes waves.
    - Ex. A rope that is fixed at one of the ends. If you apply force by flicking the rope up and then down you create waves. Without the force of you flicking the rope up and down no waves would form.
- Materials and Waves
  - A medium is anything that a wave moves through.
    - Ex. Water is the medium for an ocean wave
  - Mechanical waves are waves that transfer energy through matter.
    - Water, soil, a rope and air are all made up of matter.
- Energy and Waves
  - Waves caused by earthquakes are good examples of energy transfer.

Waves can be classified by how they move
- Transverse Waves
  - In a transfer wave the direction in which the wave travels is perpendicular to the direction of the disturbance.
    - Ex. Examples of transverse waves are ocean waves.
- Longitudinal Waves
  - In a longitudinal wave the wave travels in the same direction as the disturbance.
    - Ex. An example of a longitudinal wave is a spring that is being compressed and de-compressed.

Modifications and Other Considerations

Expert groups might be required to develop presentations on their topic or block of text. The groups then present the information to the entire class instead of to their home group. Teachers might give students information sheets with headers (and sometimes subheaders) to scaffold the recording of important information. Several teachers who routinely use Reciprocal Teaching have students in their expert group use the strategy as they develop an understanding of the information from their particular block of text. Many teachers require students to read all the text even if the information is "jigsawed." They report that it improves responsibility for the information and helps students ground concepts they were taught from the experts. Others use discussion to make sure students' developing understanding is on the right track. Some teachers use this strategy when ideas need to be extracted from large text selections but feel it isn't necessary for students to read all the material. Teachers who routinely use Jigsaw, report that students become very adept at the strategy and are able to develop relatively complex topics. One teacher shared that he assigned topics to students for homework so that the next day the students were ready to process the topics in their expert groups and then move into their home groups.

Journaling

Reading Purposes: Retrieving Information
Interpreting
Reflecting and Evaluating

Description

Journaling is a writing to learn strategy that can be used to support the processing of text. Journaling related to reading supports students in comprehending texts as important mathematical and scientific concepts are refined, described, and explained. Journaling can have multiple purposes and formats but the basic premise is that journaling provides a means for students to write on a regular basis. Interacting with text is what distinguishes journaling as a reading strategy. For example, students might be asked to summarize a block of text, explain a key concept from their reading, compare two or more concepts or ideas, list questions resulting from reading, or applying some concept or process from reading to a problem or situation. Journaling is a powerful way for students to organize and collect ideas and information and is especially effective in supporting students in learning new concepts or information, learning and using new vocabulary, and reviewing key information.

Implementing the Strategy

1. Identify a text passage and select a reading outcome. Note that several desired learning outcomes from the reading may be identified, but the journaling activity should focus on a specific outcome.

2. Present the outcome or prompt for the journaling activity. Give students a block of time to read the selected text and complete the journal task.

3. Students pair with a classmate to share their journal response (or in small groups).

4. Students revise journal responses based on peer or small group discussion using the text to check information and perspectives.

5. Journal responses are used as a springboard for further instruction.

Figure 4.16. Journaling activity on Parallel and Perpendicular Lines.

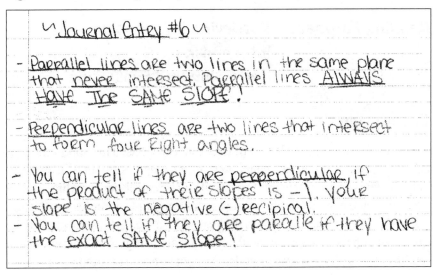

Modifications and Other Considerations

Since the journaling exercise is presented as a reading strategy, there is a strong component that involves reporting and sharing information. This helps students comprehend important information and identify problems they may have encountered in understanding the text. Allowing students to revise the information provides additional support through revisiting the text as additional or new information is embedded in the journal response. Some teachers encourage students to write additional information below the entry but require any erroneous information to be corrected in the original entry. Other types of journaling are possible; however, this particular version supports students' interaction with text.

Teachers might have students complete several journaling exercises individually during a unit of study or with several blocks of text. Teachers might modify how students use the journal entries. Peer and group writing is also a possible variation in the journaling activity (Pugalee, 2005). Students might be given more structure in journaling activities. Such structure might include a series of questions, a list of points that must be addressed, or other criteria that will help students be successful in using the strategy. As students become more adept in using the strategy, less structure can be provided until students are writing entries without any specific aids.

K-W-L

Reading Purposes: Retrieving Information
Interpreting
Reflecting and Evaluating

Description

K-W-L is an acronym that stands for a three-pronged reading approach. The **K** stands for what students already **K**now about the topic. This serves to activate their prior knowledge. The **W** stands for what students **W**ant to learn. This serves to help students identify goals thus giving a purpose for reading. The **L** stands for what was **L**earned. This provides a post-reading emphasis promoting application of ideas and concepts. The K-W-L strategy promotes students' critical thinking as they engage with text and also promotes metacognitive behaviors as students are given a structure for monitoring their progress and comprehension of the text. Students should revisit their completed K-W-L using texts and other resources to determine if any of their prior knowledge, **K**, was inaccurate. Students should compare their prior knowledge statements with what was learned, **L**, so that any erroneous ideas or misconceptions are addressed. Students should also revisit what they wanted to learn, **W**, to identify any questions that were not answered. Students should be encouraged to find answers for these questions.

Implementing the Strategy

1. Select the text that is the focus of the activity. Identify the primary topic of the reading. This identification might be one or more words or a phrase.

2. Students take some time to complete the **K**now part of the strategy by listing or describing what they already know about the topic.

3. Students identify what they **W**ant to learn about the topic completing the **W** part of the strategy.

4. Students complete the **L**earn part of the strategy during and after the reading by identifying what they learned from the text.

5. Students compare the three components of the strategy paying particular attention to correcting any misconceptions in their understanding of the topic and identifying questions that remain unanswered or new questions that arise from the reading.

Figure 4.17. K-W-L on Graphing.

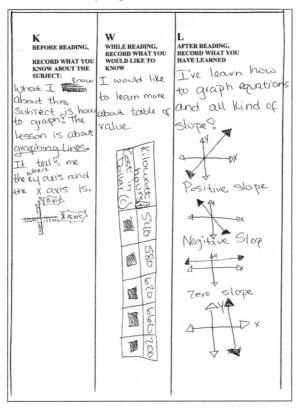

Modifications and Other Considerations

The last step in implementation is sometimes not emphasized when implementing the strategy, probably because the K-W-L chart does not include a grid for explicitly dealing with this type of follow-up. Many reading specialists stress how important it is for students to revisit their ideas and compare them to knowledge and understanding gained through exploring the text. Through this process, students are required to revisit ideas that may need to be restructured in their schema for the particular topic. Identification of questions not answered or new questions that have been identified is an important tool in motivating students to do additional reading. Some teachers accomplish the last step by having students write a short summary that includes these points. Other teachers use class discussion to draw attention to connections between prior knowledge (**K**), information that students want to know about the topic (**W**), and what was learned from the reading (**L**).

One secondary geometry teacher shared an interesting modification to the strategy. He emphasized that frequently students lack background knowledge for a particular topic. He uses what he refers to as a K-L-W. First, students read a text passage and use that information to identify what they know (**K**). Instruction follows, sometimes teacher directed and sometimes more discovery oriented. Students use the **L** part of the strategy to identify what was clarified or what they learned from the instruction. The **W** part of the strategy focuses on what questions students have about the topic or what is unclear to them. The teacher shares how effective the strategy modification is for his students. He added that he will sometimes use the traditional method if he feels that students will have some knowledge of the topic. He points out that modifying the strategy helped because students would often become frustrated if they felt they were unsuccessful in identifying prior knowledge that was relevant to the topic.

A middle grades science teacher added that she has a huge K-W-L chart on a bulletin board that has been laminated so that it can be written on and later erased and reused. For each of the three components, students are given an opportunity to complete the chart. They then draw a line and add information that emerges from the class discussion. She engages the entire class in completing a class composite for each section. The **K** is a class-wide brainstorming on what students know about the topic and the generation of questions that students would like to know more about (the **W**). Students individually record what they learn during the reading and review it after they have completed the entire passage. Class discussion provides an opportunity for students to share information that is added to the class K-W-L chart.

Learning Log

Reading Purposes: Retrieving Information
                   Interpreting
                   Reflecting and Evaluating

Description

Learning logs have traditionally been used with literature but can be used effectively with all forms of text. Learning logs provide an opportunity for students to reflect on what they are learning, what they find interesting, or difficulties they encountered in their reading. Teachers recommend a separate notebook for students to use as their learning log. Some use the term learning log and journal interchangeably. Learning logs provide a tool for students to make direct relationships between text and their understanding. The strategy provides a reflective tool for students so that they are more aware of their own learning and are able to identify successes and obstacles in that process. Used in this way, learning logs provide an excellent record for teachers to consider in identifying strengths and weaknesses in students' understanding relative to a particular topic or text.

Implementing the Strategy

1. During reading, identify several major points from the text, or record an idea or question that you have about the topic. Record these ideas in one column (or at the top of a page).

2. Use information from the text to respond to the major points.

3. Reflect on the learning process and identify difficulties (and how they were handled) or successes. Summarize this reflection at the end of the log.

Figure 4.18. Learning Log on Function of Cells.

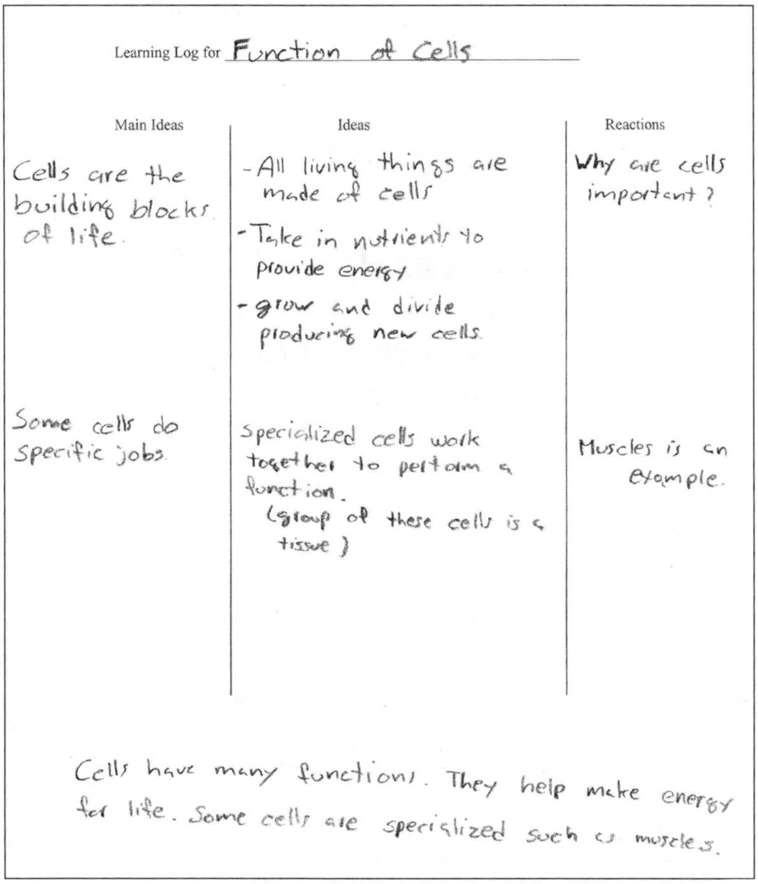

Learning Log for Function of Cells

Main Ideas	Ideas	Reactions
Cells are the building blocks of life.	- All living things are made of cells - Take in nutrients to provide energy - grow and divide producing new cells.	Why are cells important ?
Some cells do specific jobs.	Specialized cells work together to perform a function. (group of these cells is a tissue)	Muscles is an example.

Cells have many functions. They help make energy for life. Some cells are specialized such as muscles.

Modifications and Other Considerations

Learning logs can be modified in multiple ways. The instructions given here are a basic framework. Logs, in general, are modifications of journals. A modification of a learning log is a reading log which makes an explicit association between text and a students' understanding of the text. A reading log usually consists of two columns: a column for short phrases or ideas from the text (sometimes verbatim), or observations and analysis from the text, and a separate column for the student's ideas about what they know or understand about the idea including comments, questions, and evaluations about their learning.

Other teachers might use reading logs differently, such as responding to a particular prompt or exercise; however, the idea is that the experience requires students to analyze text in their response. Some teachers use the learning logs as a modified dialogue journal by providing periodic feedback or by hav-ing students respond to another student's entry. Some teachers provide more structure for the ideas to be addressed in the learning log. A mathematics teacher has students write learning log entries based on something they didn't understand in the reading. After the lesson is developed students revisit their entry and tell how their learning changed and what helped clarify their thinking.

List-Classify-Identify

Reading Purposes: Retrieving Information
Interpreting

Description

This activity is designed for use before beginning a new chapter or unit of study. The strategy helps students examine relationships among key math-ematical and scientific concepts. Students use their own vocabulary as they offer words that are associated with a concept or idea; therefore, prior knowl-edge is activated before students deal with a topic or concept in their studies. The activity also provides a tool for the teacher to assess students' current level of understanding and to identify any erroneous learning or misconceptions.

Implementing the Strategy

1. Teacher identifies a concept term and poses it to students. The term may be written on the board or other public display area.

2. Students generate words or phrases that they associate with the word(s).

3. Encourage students to continue offering ideas until they have gen-erated a list of 15 to 30 words.

4. Ask students (in pairs, groups, or as a whole class) to group words that have something in common. Emphasize that the categories should identify an important relationship among the words.

5. Identify the word or phrase that best describes the relationship.

6. Engage students in providing a rationale for how they grouped the words together.

7. Ask students to explain their rationale for grouping words together.

8. Through the discussion, emphasize key concepts that will be the focus of the unit or chapter.

Figure 4.19. List-Classify-Identify for Classifying Triangles.

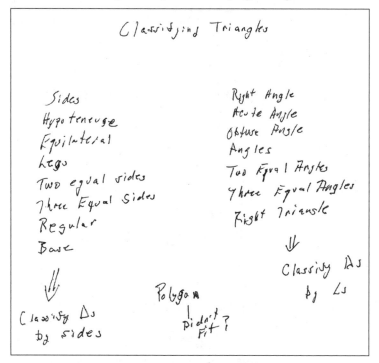

Modifications and Other Considerations

In the example, students completed their work after brainstorming words about classifying triangles. The words were written on the board. In this case, notice that the student pair grouped words into two categories: one for classifying triangles by angles and one for classifying triangles by sides. One word, polygon, didn't really fit either category. The teacher had deliberately omitted several words offered by the students because they were not clearly associated with classification of triangles. One such word was parallel which isn't related to triangles. The teacher explained why the offered words wouldn't fit and gave students a chance to expand on their choices. This activity was an introduction to several lessons that would explore classifying triangles and finding missing angle and side measurements for various types of triangles.

This strategy is especially powerful if different groups take the words and classify them into groups. The discussion of the results will likely reveal different ways of thinking about the words. This opens up opportunities to help develop a deeper understanding of the concepts by exploring different perceptions of important relationships.

Teachers sometimes give students a collection of words related to a topic instead of having students generate them. This helps the teacher identify what students know about particular words and the concepts embedded within those meanings. Some might modify the activity by making it a post-reading exercise. This changes the focus to what was learned from the lesson instead of a focus on using the activity to activate prior knowledge.

Parallel Notes

Reading Purposes: Retrieving Information
Interpreting
Reflecting and Evaluating

Description

Parallel notes is a structured note-taking strategy. It is designed to assist students in using organizational features of text in taking notes that facilitate the comprehension of key concepts and essential ideas that **parallel** the structure of the text. The structure of the text guides students in taking notes that respond to those features. In science and mathematics, these features frequently include an example or diagram. In general, parallel notes assist students by helping them identify how the structure of their text is used to develop ideas or concepts. Teachers initially identify the features that characterize a piece of text until students become familiar with the approach and can identify the organizational structure on their own.

Implementing the Strategy

1. Identify the organization of the text being considered.
2. Use these features as a way of organizing and identifying key information from the text.
3. Process or synthesize the information from the notes. The most common method is for students to use their notes as the lesson is

developed, comparing their information with key ideas presented in class.

Figure 4.20. Parallel Notes on Building Plans.

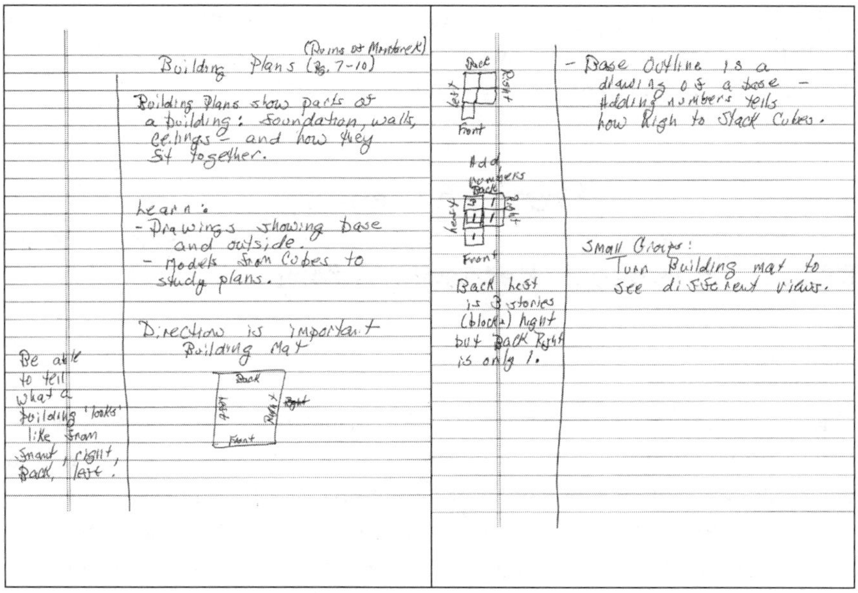

Modifications and Other Considerations

In this example, the student first records details and concepts from individual reading. The column on the left is used to make follow-up notes during the further development of the lesson. The student's notes include information on what he should be able to do after the lesson (tell what a building looks like from different perspectives). The notes in the left column also include the example used in small groups as the teacher engaged students in exploring the goals of the lesson.

Reading researchers sometimes refer to seven common organizational patterns (see Jones, Palincsar, Ogle, & Carr, 1987): chronological (order of events), compare/contrast (features of two or more things focusing on similarities and/or differences), concept/definition (bid ideas and details), description (details that specify a concept, idea, or process), episode (before/during/after or who, what, when, where, why, and how), generalization (broad idea or concept and the supporting details), and process/cause-effect (events and results). The strategy frequently involves the use of graphic organizers to assist students in following the process. Some teachers prefer to identify the structural elements of the text without a visual organizational tool. These teachers stress that while graphic organizers are useful, they sometimes over-simplify the text resulting in students missing key ideas or concepts.

It is important that students have an opportunity to process their notes. Some ways that teachers recommend is to have students work collaboratively, use the notes to produce a written product such as a summary paragraph, or apply the information in some way that shows an understanding of the processes identified in the notes.

Prefix Mastery

Reading Purposes: Retrieving Information

Description

Understanding prefixes is an important skill in decoding text. Students should understand what prefixes are and how they contribute to the meaning of a word. Students who know how prefixes change the meaning of root words will be able to negotiate new terms more easily. In addition, students need experience using their knowledge of prefixes to determine the meaning of important mathematical and scientific terms. This strategy is intended to be a long-term strategy allowing addition of relevant terms as the school year progresses.

Implementing the Strategy

1. Give students the word or words containing a specific prefix.

2. Allow students to develop an understanding of the prefix's meaning by having them list other words that have the prefix and speculating on how the prefix changes the meaning of the words.

3. Give students or have students share their thinking with the end result being that the students have the correct understanding of the meaning of the prefix and the meaning of the word(s) being studied.

4. The teacher and/or the students can add additional math or science terms being studied that have the prefix.

5. Add additional terms to the table as they are encountered in various lessons. Using a similar process of having students think about the prefix and reflecting on what they may know about its meaning.

Table 4.2. Common Mathematics Prefixes.

Prefix	Meaning	Example
Bi	Two	Bisect: Cut into two equal parts
Centi	One hundred	Centigram: 100 grams
Circum	Around	Circumference: Distance around a circle
Con, Com	With or together	Con
Dia	Through	Diameter: Segment through the center of a circle
Equi	Equal	Equivalent: Being equal or the same
Mid	Half	Midpoint: Point half way between endpoints
Para	Beside	Parallel: Being of equal distance
Penta	Five	Pentagon: Polygon with five sides
Peri	Around or About	Perimeter: Distance around a figure
Poly	Many	Polygon: Figure with
Quad	Four	Quadrilateral: Figure with 4 sides
Semi	Half	Semicircle: half of a circle
Tri	Three	Triangle: Polygon with three sides

Modifications and Other Considerations

This strategy can easily be modified to deal with suffixes and even root words. Suffixes such as -ity (state of being), -al (of or pertaining to), -form (shape or appearance) that are common in mathematics and/or science can help students figure out the meaning of words. An emphasis on root words can build helpful word recognition skills, particularly for science. Roots such as cardio (heart), pod (foot), and hydro (water) are used frequently in scientific terms.

Teachers extend the emphasis on prefixes by using the list for a quick card sort game. Each prefix requires two index cards. The prefix is written on one card and the meaning on another. Students can use the cards individually, in pairs, or small groups to review the meaning of the prefixes. New prefixes can be added at any time. Teachers also sometimes use the cards as a warm-up by calling out the prefix or meaning and seeing how many students know the answer.

Several teachers construct a large table on poster paper or on a bulletin board so that they can add prefixes as they are being studied. One teacher gave the students a copy of a table at the beginning of the year and had the students tape it into their notebooks. Students added prefixes as they were encountered in the lessons. Teachers may add an additional column that allows for a summary containing important information or a diagram/model pertaining to a word.

Number and metric prefixes are important in reading mathematical and scientific information. Understanding these prefixes can help students decode the meanings of important terms. Tables 4.3 and 4.4 contain number and metric prefixes that students might encounter in their readings and explorations in mathematics and science.

Table 4.3.   **Common Number Prefixes.**

Prefix	Prefix meaning
uni-	1
mono-	1
bi-	2
tri-	3
quadr-	4
quint-	5
penta-	5
hex-	6
sex-	6
hept-	7
sept-	7
octo-	8
novem-	9
deka- or deca-	10
cent-	hundred
hecto-	hundred
milli-	thousand
kilo-	thousand
mega-	million
giga-	billion

Table 4.4. Common Metric Prefixes.

Prefix	Meaning	Exponential Form
giga	1,000,000,000	10^9
mega	1,000,000	10^6
kilo	1,000	10^3
hecto	100	10^2
deca	10	10^1
	1	10^0
deci	0.1	10^{-1}
centi	0.01	10^{-2}
milli	0.001	10^{-3}
micro	0.000001	10^{-6}
nano	0.000000001	10^{-9}

Presentation Project

Reading Purposes: Retrieving Information
Interpreting
Reflecting and Evaluating

Description

Technology provides opportunities for students to develop computer skills while working with mathematics and science texts. Tools such as Kid Pix®, PowerPoint, Inspiration, and other programs motivate students to explore texts. The second grade science example illustrated below was created with Kid Pix®. Through developing a presentation, students develop skills in formulating questions, utilizing various text forms to locate information, comprehending text, synthesizing and reflecting on content, and extending written and spoken communication skills. Presentation activities assist students in developing skills, locating information in texts, summarizing and synthesizing multiple ideas and facts, and expressing concepts and information precisely and informatively.

Implementing the Strategy

1. Identify a topic or concept that students will research.

2. Formulate questions or topics for exploration. Students can come up with questions in small groups based on their reading or the teacher can provide students with appropriate questions or topics.

3. Students, individually or in small cooperative groups, complete the map based on their reading.

4. A short summary slide is created based on the information from the concept map.

5. If students are working on different topics, they can present their information to the class or the information can be assembled in some form to be used for whole class instruction.

Figure 4.21. Students' Summary on Diplodocus.

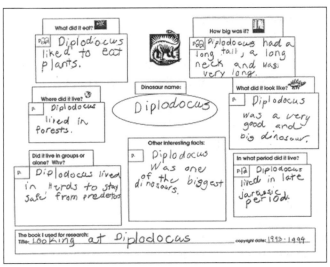

Figure 4.22. Students' Slide for Diplodocus Reading Activity.

Modifications and Other Considerations

Presentation projects can be carried out without using technology. Classroom teachers without access to these tools can provide students with paper copies of frameworks or concept maps. The important element in presentation projects is that students are interacting with text to broaden their understanding of key concepts. Regardless of the tools used, the presentation should contain a summary organization such as the map on the Diplodocus in the example along with a summary slide that puts those key ideas together. Students can become creative in illustrating their work and might be encouraged to include voice files, music, and other media formats into their presentations.

This strategy works especially well with expository text. In the example, the students were using trade books to learn more about various dinosaurs. Trade books such as the one used in this example are effective in developing reading skills and motivating students to read. Trade books are available for students at the different grade and ability levels. The presentation project strategy can be used with other text forms including Internet-based information sources, newspapers and magazines, textbooks, and so on. As students research, they are developing learning that spans multiple literacies as well as extending their understanding of mathematical and scientific concepts in deep and relevant ways.

Problem Solving Seven

Reading Purposes: Retrieving Information
Interpreting
Reflecting and Evaluating

Description

Problem Solving Seven is a strategy that helps students organize information, develop and execute a plan, and reflect on their process. While it is especially suited for mathematics, the process can be applied to other situations that involve using information to approach a problem. Polya's approach to problem solving (1957) provides a framework for the strategy. Polya emphasized four key steps in the problem-solving process: understand the problem, devise a plan, carry out the plan, and look back. Problem Solving Seven is designed to help students focus on understanding how textual information guides this process. The strategy underscores recognizing the concepts related to key vocabulary. More importantly, the strategy helps students reflect as they move back and forth between their problem solving plan and the text.

Implementing the Strategy

1. Read the problem carefully and identify key words that are crucial to the plan. It is helpful to underline or highlight these key words.

2. Associate actions to the key words identified in the problem. This will help in devising a plan.

3. Restate what the problem is asking for—the end product.

4. Recall similar problems. Record what you remember about the problem.

5. Devise a plan based on the information. Identify the information in the text that supports the plan.

6. Execute the plan.

7. Verify the outcome. Review what the problem is asking for and tell how your outcome is reasonable.

Figure 4.23. Problem Solving Seven on Angles in a Triangle.

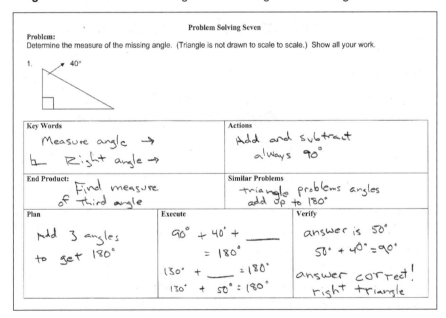

Modifications and Other Considerations

In the above example, the teacher provided a grid that emphasized the seven stages of the problem-solving process. The teacher also includes the problem printed on the sheet. She has found that this helps students focus on applying the steps in solving the problem. This problem is indicative of the "entry-level" problems used when applying new concepts. Notice the importance of the diagram in using the plan. The student notes that the triangle is a right triangle using key words that require an interpretation that the angle measure is 90 degrees. When students are taught to deal with diagrams as part of understanding a text passage, they develop the capabilities of extracting key information from such diagrams and illustrations.

Students may need review of key clue words that are appropriate for their course. For example, average and distribute are key words that indicate division. As students begin to organize information, encourage them to draw a picture or diagram; make a chart, graph, table, or list and sort data to assist in analysis of characteristics and properties of the data set. Encourage students to label units of measurement and identify variables and unknown quantities. These are good thinking practices regardless of whether students are dealing with a problem in mathematics or science. It is also helpful to sometimes give students a similar problem that they have seen in class when they get to step 4. Some teachers give them several problems and ask the students to use the one that is most like the current problem.

When developing this strategy and implementing it with students, they sometimes complained that the plan was too much work and they didn't need to do all the extra recording. As with any strategy, it should be modified as students become adept in using it. Strategy use should become "second nature" as students internalize the underlying processes. The use of the strategy (or any tool, for that matter) requires students to show that they are following the procedure until they can demonstrate that they no longer need to show the steps. Problem Solving Seven should become a cognitive tool that students begin to use with less structure. The key is that they are employing the thinking framework behind the strategy.

Question Answer Response (QAR)

Reading Purposes: Retrieving Information
Interpreting
Reflecting and Evaluating

Description

Question Answer Response or QAR is a reading strategy that helps students understand types of comprehension questions according to their approach to find the information. Students need experiences developing approaches to questions based on the amount of cognitive engagement necessary to find the information. There are three types of questions. The first type is **Right There** questions where the answer is found explicitly in the text. The second type is **Think and Search** questions which are textually implicit or the answer is implied in the text. These two types are often referred to as "in the book" while the third type is "in my head." The third type is **On My Own** questions where the answer depends on the student's background knowledge or prior experience and the answer is not found in the text. The third type of question usually involves an application of information or concepts. This strategy helps students monitor their comprehension of text, provides a purpose for reading, improves students' assessment of their comprehension, and encourages critical thinking.

Implementing the Strategy

1. Explain the three categories of questions as an introduction to the QAR method or review as necessary when using the strategy. Model as necessary to demonstrate how to think about answering each of the question types.

2. Provide the reading selection and the set of questions for each of the categories. The teacher constructs the question based on learning targets or goals for the lesson. Stress to students that they should think about how they approach each type of question type so that they internalize how those approaches differ based on how the answer is found from the reading.

3. Working in small groups or individually, have students answer the questions in each of the categories.

4. As a class, pair, or group activity compare products assimilating information and concepts into a final class or group product. Teachers might ask students to work on one category of questions at a time so that the process is more evident.

Figure 4.24. Question Answer Relationship (QAR from sixth grade mathematics)

Right There

- What is an inequality?
- What does an open dot on the number line mean?
- What is the sign for greater than? Less than?

Think and Search

- Write an inequality for n is less than 6.
- Show an example of a false inequality.
- How would x > 10 be graphed?

On My Own

- Give an example of a real-life inequality.
- Compare and contrast inequalities and equalities.
- Find several values that are true for 3 > x > 10. What values would make this false?

Modifications and Other Considerations

This strategy is also appropriate as a whole class. This is especially effective when students are becoming familiar with the different types of questions and the teacher is helping them understand how to use the question types to understand how to find the information in the text. Some modifications include having the students classify the questions as one of the three types as they read the selection. This helps them apply the classification system. Another modification involves having students write questions for each of the categories as part of reading a selection. Students could exchange questions or the teacher could compile questions to use as part of a review or discussion of the material.

Some QAR models have four types of questions. **Author and You** is a fourth type that involves questions that the text provokes. These types of questions require the reader to formulate their own ideas or opinions. These questions often involve developing ideas about the implications or suggestions of the author. For example, "The passage implies that" would be a likely stem for this type of question. These types of questions are more prevalent in narrative text or in opinion or argumentative pieces, and are somewhat less common question types for science and mathematics.

Role, Audience, Format, Topic (RAFT)

Reading Purposes: Retrieving Information
                       Interpreting
                       Reflecting and Evaluating

Description

This strategy connects reading and writing. The strategy requires students to apply information in generating a new means of demonstrating their understanding of the text. Teachers like the strategy because of the high level of flexibility in the implementation. **R** stands for role where students identify the role they will take in the new product. Roles might include author, character, artists, performer, expert, mathematician, scientist, and so on. **A** stands for the audience for the product. Potential audiences might include oneself, other students, teacher(s), parents, the community, and other known or unknown individuals or groups. **F** stands for format or the best way to present the information. Students might write a report, a play, a poem, a song, construct a presentation, and so on. **T** stands for topic or the who, what, when, where, why and/or how that will be the focus of the product. The possible products include a range of written formats including creative or report writing, letters, memos, plays, poems, presentations, and other forms of writing. The strategy requires students to consider key questions about the writing process including audience and format. The strategy can be used in various contexts and is especially suited for making connections between language arts and mathematics or science.

Implementing the Strategy

1.   Students select the role, audience, format, and topic for their product. Students might work in pairs or small groups but will need to agree on the specific components.

2. Students use text(s) and other print or media materials to guide them in creating their product. The product must be different from the original format.

3. Students review their product paying particular attention to the information that is used to develop the topic. Students must also evaluate the appropriateness of the information and format for the intended audience.

4. Students present their final products for dissemination.

Figure 4.25. RAFT Product on Form of Linear Equations.

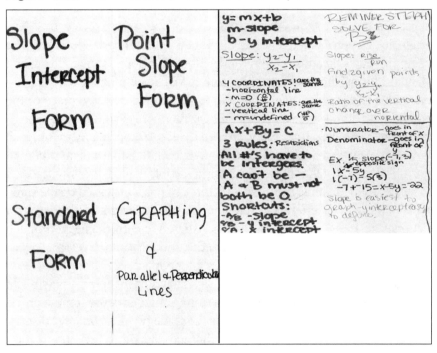

Modifications and Other Considerations

In the example presented, the teacher developed several pieces for which students could develop materials on writing linear equations: introduction of the concept, practice and problem solving, or review. Students worked on their parts of several lessons focused on writing linear equations. Students worked in pairs once their responsibilities had been specified. They used the text and other material to explore central ideas. The teacher served as facilitator making sure students had no difficulty with ideas and concepts they encountered in their readings. The above example comes from one of the pairs of students who decided to focus on review of material. The pairs created materials that they thought would be helpful in reviewing key concepts. The example shows the outside cover of one of three review booklets. Each block when opened revealed information on that particular topic. The example shows the information for the blocks on slope-intercept form and standard form. The teacher emphasized RAFT in helping the students expand ideas about what they might do to fulfill their selected piece of the assignment. Some students created PowerPoint presentations, made posters, created games, wrote a play, created a role-play, and many more.

The strategy can be used effectively with both fictional and nonfictional texts. The appeal of the strategy is that it is flexible enough to allow students to be creative. Students can create a variety of products based on the same topic or piece of literature. One elementary teacher uses RAFT along with fictional text by asking students to demonstrate a scientific concept from the reading. RAFT encourages them to identify a concept and develop a way of communicating how the concept relates to important ideas from science.

Dissemination of RAFT products can be handled in multiple ways. The RAFT framework can be used as prewriting activities or to promote group and class discussion. Teachers who use the strategy often will have display boards for showing the products or encourage students to keep a collection of their products over a unit or grading period. Some RAFTs may require role playing or other physical activity to carry them out. These provide excellent opportunities to extend instructional practices to provide a wide range of opportunities as well as multiple ways to represent concepts. As such, RAFT is an excellent tool for differentiation of instruction.

Reciprocal Teaching

Reading Purposes: Retrieving Information
Interpreting
Reflecting and Evaluating

Description

Reciprocal teaching is a strategy that engages the teacher and students in dialogue regarding segments of text with the dialogue structured by the use of four strategies designed to support reading comprehension: summarizing, question generating, clarifying, and predicting (Palincsar, 1986). The students and teacher share responsibility in this strategy through turn-taking. The four components do not necessarily have to be done in order, but using all four is essential in using the strategy to support text comprehension. Before using the strategy, students need instruction and practice using the four strategies. As students become more proficient, the teacher takes less of a direct role and monitors/facilitates the peer-driven learning.

- **Summarizing** involves identifying and integrating important information from the reading. Students may begin at the sentence and paragraph level when they first begin use of the strategy but move to blocks of text as they become more familiar with the process.

- **Question generating** develops inference making by having students generate questions. First, students identify the information that is significant enough for basing a question. Next, students pose this information in a question and self-test to make sure the question/response is substantive and they can answer the question.

- **Clarifying** assists students in identifying where and why the text doesn't make sense. Students' attention may be called to new vocabulary, unclear references to concepts or ideas, and unfamiliar concepts or ideas for which they have little background knowledge. This process encourages students to reread, look for additional information, and seek help from others.

- **Predicting** engages students in hypothesizing what will be discussed next in the text. This supports a purpose for reading, activates prior knowledge, and helps students effectively use text structures (headings, subheadings, and questions embedded in the text).

Implementing the Strategy

1. Form student groups of four.

2. Give each student a notecard or piece of paper with one of the four components written on it. This identifies the students' role later in the discussions.

3. Students read selected text. They should be encouraged to take notes to help them prepare for the discussion.

4. After allowing adequate time, the groups are instructed to discuss the passage. This begins with the Summarizer providing an overview of key ideas in the reading. The Questioner then poses questions and guides the group in generating answers. The Clarifier identifies confusing points and the Predictor will offer predictions about what might come next in the text.

5. The members of the group rotate cards and move to the next passage of text. The students now read the next passage paying attention to their new roles. The process is repeated until the entire reading selection is completed.

6. The process should include some type of closure activity such as whole class discussion or group reporting. Many teachers also provide a four column sheet with the components as headers for students to record their work.

Figure 4.26. Student Record Sheet for Reciprocal Teaching Activity on Ecosystems

Summarize	Question	Clarify	Predict
Puting the wolfs back in Yellowstone Natinal park.	Is puting the wolfs back in yellowsfom park a good thing.	Yes its good because cotten wood trees are coming back.	Predict that the wolfs wrll be there for a long time.
How the wolfs are helping the ecosysfem.	Why wasent the ecosystem good when just the Elk were there.	The ecostem wasent good becaus all the elk were eating all the plants.	that the wolfs keep on helping the ecosystem.
How a little bet of the Elk pouple lation went down.	How did the poupulation of the elk went down.	Wof are carnivors they like to eat meat.	Predict that the wolfs will keep on eating the Elk.

Modifications and Other Considerations

It is important for students to use effective strategies in independent reading. Teachers who use reciprocal teaching will sometimes have students use the process for independent reading and record their ideas. Students might later work in pairs to discuss how they used the strategies to understand the text. Teachers may also want to develop writing activities such as journals or logs to summarize discussions from several blocks of text developed through reciprocal teaching. When using the strategy in mathematics, it is important for students to understand that examples, tables, diagrams, and other features are important parts of the text.

Teachers find it necessary to monitor the discussions of students to determine if the strategy is being used appropriately and to what degree it is helping students. Teachers can provide feedback and scaffolding depending on how the students are performing. Asking students to write questions and summaries provides a mechanism for the teacher to more carefully monitor students' progress and use of the strategy. As students take on more responsibility for monitoring and assessing their own learning, the level of monitoring can be adjusted. Students still benefit, however, from written records of their thinking since writing promotes reflection and ownership.

Semantic Feature Analysis

Reading Purposes: Retrieving Information

                            Interpreting

                            Reflecting and Evaluating

Description

Extending understanding of concepts frequently requires students to compare and contrast features, properties, events, and so on. Semantic feature analysis supports this type of thinking by focusing attention on how ideas, concepts, or objects are related. Through this process, students develop a more in-depth understanding but also see how things are connected. The activity also draws on students' prior knowledge of the concepts, ideas, or items being considered. The strategy helps clarify vocabulary related to the concept or topic, reinforces concept understanding, and helps them think of how ideas are similar and/or different. A grid arrangement is used to give students a visual aid for comparing. The strategy can be used as a whole class activity, in pairs, groups, or individually.

Implementing the Strategy

1. Identify the topic to be considered.

2. Make a list of components or elements related to the topic. These will be listed across the top of the columns in the grid.

3. Make a list of features, characteristics, or criteria that some of the components or elements might possess. These will be listed in the left column of the grid.

4. The grid is completed by placing a + in cells where the component/ element along the top of the column possess the characteristic, feature or criterion listed for that row of the grid, or a – if the feature is absent.

5. Consider the results. Which columns are similar? Rows? What features do these components have and how would they be described or labeled?

Figure 4.27. Semantic Feature Analysis for Number Systems.

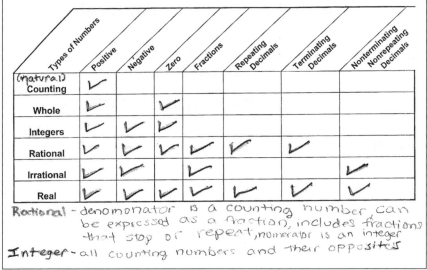

Modifications and Other Considerations

Some students find it less distracting if their grid only contains +s. This helps them quickly process which components have which features, characteristics, and/or criteria. Students may place question marks in the grids if they are not sure so that they can go back and fill those cells in with the appropriate response.

Teachers who use this as a during-reading activity provide the labels for the rows and columns. Students complete the grid while they are reading. Students can help identify the headings for the columns and rows as an after reading activity. Once students become familiar with the strategy, teachers may have students construct their own semantic feature analysis. Students can compare these to others or modify them as a result of discussion or additional instruction.

Some teachers sometimes use the semantic feature analysis throughout a unit of instruction. A copy on a transparency or a grid on poster paper can provide a place for recording information as it is encountered in the lessons. Additional rows and columns can be added as necessary. Students can update their own copy as the one in class is developed. This provides an ongoing way of comparing and contrasting ideas or things over several lessons. Semantic feature analysis grids also provide useful tools for reviewing information. Some teachers include grid to be filled in on assessments or they may provide partially completed examples that require listing missing labels for the columns or rows in addition to filling in individual grids as necessary.

Semantic Mapping

Reading Purposes: Retrieving Information
Interpreting

Description

Semantic mapping is an activity that engages students in identifying important ideas or concepts and how they fit together. Semantic mapping also helps students improve their understanding of vocabulary by considering how words are connected. They provide a means of connecting what is already known with new concepts and ideas. In addition, students must identify multiple relationships between those concepts and ideas. These associations may be related to classes or the order of things related to the concept, properties, or the attributes that describe and define the concepts, or examples which provide exemplars or models of the concept. Maps generally have three components: (1) a central question, word, or phrase that is the focus of the map; (2) strands or subordinate ideas that clarify, explain, or describe the main idea; and (3) supporting information that includes generalizations, details, inferences, and examples that provide additional information that helps distinguish between strands.

Implementing the Strategy

1. Identify a word or phrase central to the topic.

2. Students generate as many words as possible that relate to the word or phrase.

3. Students write the words in categories.

4. Students label the categories using a word or phrase that best describes the association between central features or characteristics of the words.

5. Students use this list to construct a map or diagram.

6. Discussion focuses on identifying the meanings and uses of the words, making conclusions about how the words are related and summarizing and identifying central concepts and themes.

Figure 4.28. Semantic Web on Bats

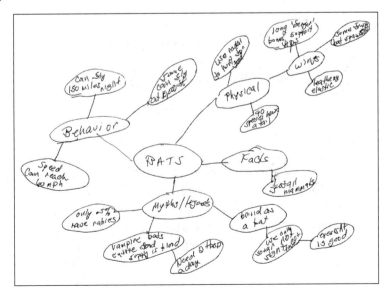

Modifications and Other Considerations

The above example is one type of modification for this strategy. Students were given the major categories for the topic of "Bats" before reading a short article that used a lot of callouts, similar to those text bubbles used in comics, and visuals to provide information about bats. The short piece was found in a travel magazine and the science teacher used it to stimulate interest as part of a lesson on mammals which took place close to Halloween. The students were given the major categories: physical, behavior, facts, and myths/legends. After reading, the class completed the semantic web as part of a discussion of their reading. This example is a student copy of the class product.

A common modification is for the teacher to provide students with a list of words or phrases that are applicable to the lesson and ask students to classify them, come up with the appropriate label and construct a map. Teachers report that this is helpful if there are concepts or ideas that may be challenging for students or if there is a need to make sure students focus on central topics or ideas. Some teachers give students the labels to be used and have students complete the map by providing words for the labels. Some teachers also give students a map that is completed except for the labels of the concepts. The level of information given depends on the purpose the teacher has for the activity and the students' level of proficiency in using the strategy.

Teachers, who use semantic mapping for several related lessons, may have students repeat the steps for each lesson. Additionally, students make connections between the two maps by using showing how various groupings are related. For example, a label and the set of words associated with that label may be included under the label and associated words for a previous map. This helps students see how lessons are connected and gives them a frame for organizing the information cognitively.

Sentence Cards

Reading Purposes: Retrieving Information
Interpreting

Description

Sentence cards reinforce vocabulary development and help students recognize and understand key ideas and information. Sentence cards can be developed to meet multiple instructional goals. Various types of sentence cards are available commercially. Sentence cards differ from flash cards in that flash cards focus more on skill development and recall. On a basic level, sentence cards can be used for word recognition. As presented here, sentence cards require students to locate or recall information and details from text in order to complete the sentence cards. Sentence cards might also involve diagrams, illustrations, models, or examples that relate to the reading.

Implementing the Strategy

1. The teacher identifies key ideas, concepts, and definitions from the reading.

2. The teacher writes statements that correspond to these key pieces of information.

3. The teacher removes one or more words from the sentence.

4. These sentences are written on sentence cards (or strips). (Cards with the missing words might also be constructed at this time.)

5. Students complete the sentences by giving the missing information or displaying the appropriate card(s).

Figure 4.29. Sentence Cards for Transformations.

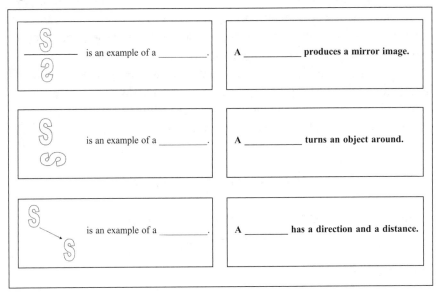

Modifications and Other Considerations

The above example shows cards that are for one lesson on transformations that was part of a fifth grade math lesson. The teacher constructed cards for each lesson and used them initially as a pre-reading activity to give students a purpose for reading. After reading and/or instruction on the lesson, students complete the sentence cards either individually or in pairs, groups, or as a whole class. The cards then become part of a larger set that could be used for reviewing concepts. Multiple sets of cards were made so the entire class can use them and are laminated to make them sturdy and durable. The cards can be reused for several years.

One modification involves constructing completed sentence cards to be used in true/false exercises. Students working in pairs or small groups take turns selecting a sentence from the stack. If the stacks of cards which the pairs or groups are using is different, it is important to keep up with which stack was used so that all students have an opportunity to consider all the cards. The object is to correctly identify the statement as true or false and provide a short justification. If the response is correct, the student giving the answer keeps the card; otherwise, it goes to the bottom of the stack. The student with the most sentence cards at the end of the activity wins the round. Teachers who use the strategy in this way save cards from year to year. They also have students construct cards as part of an assignment.

Another popular modification is to have sentence cards that give a concept, idea, or definition. Matching cards are constructed that contain a label, diagram, or word corresponding to the information conveyed by the sentence card. Teachers might hold up the sentence card and ask which student has the match—or have students record their response on a blank card and hold it up. Teachers also reverse the procedure and have students hold up the sentence that matches the label, diagram, or word on their card. This can also be done in small groups or pairs. It is an excellent review activity and motivates students to recall information and ideas from their reading. One teacher also shared using sentence cards in a center where each student recorded the missing words for each of the sentence cards. Another teacher constructs five sentence cards (she uses sentence strips) and displays them before students read a section of text. Students complete the sentence cards as they read. Going over the answers for the cards provides a way of engaging the lesson and helps uncover difficulties in understanding key ideas and concepts. Students can also be asked to develop sentence cards after reading as a way of identifying and reviewing key ideas and concepts.

Sequence Frame

Reading Purposes: Retrieving Information
Interpreting
Reflecting and Evaluating

Description

Sequence Frame is a strategy that emphasizes the order of events or actions. It helps students focus on important concepts while emphasizing that there is a sequence that underscores the information. Students frequently experience some difficulty identifying the sequence of events or actions from reading. Texts may often discuss a sequence of events without explicit attention that an order of events is being described. Mathematics contains many instances of procedures that require a sequence of specific actions. Science contains numerous examples of reactions or cycles where sequencing of events is important. The grids help students move from one action or event to the next. Additional grids or blocks can be added as necessary. When arrows are used, they are not intended to demonstrate a linear progression of the activities or events, but only to show that order is emphasized.

Implementing the Strategy

1. A concept is identified that requires the development of a sequence of events or actions.

2. As students read, they describe the events or actions recording their information in the grids.

3. Students identify whether the sequence ends; continues with a rule, pattern, or series of reactions; or repeats as a cycle.

Figure 4.30. Sequence Frame for the Water Cycle.

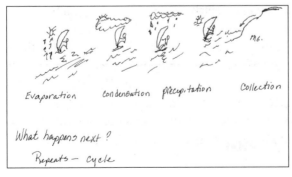

Evaporation   Condensation   Precipitation   Collection

What happens next?
Repeats – cycle

Modifications and Other Considerations

It is sometimes important that students do not develop ideas that a sequence is linear and that going back and forth between steps is somehow a "bad idea." For example, mathematical procedures such as adding two fractions may have an order, but students should be encouraged to move back and forth between steps to monitor their understanding instead of thinking that progression to one step means the previous step is complete and shouldn't be revisited. Teachers may select to have two-directional arrows between the grids to demonstrate that while the sequence is important, it is also important to reflect back on the previous actions even when performing a different step in the sequence.

The activity also asks students to reflect on whether the sequence ends, continues, or repeats. Many sequences end after a certain number of steps, such as solving a specific math problem. Some events continue by following a rule, such as a pattern in mathematics. Other events are cyclic in nature, such as the water cycle or the seasons. The sequence frame strategy adds this additional element so that students become aware of the complexity of how events are related to other events.

Story Card

Reading Purposes: Retrieving Information
Interpreting
Reflecting and Evaluating

Description

The strategy reinforces reading and writing. As students write their own story situations, they are demonstrating knowledge of problem situations. This is an important component of effective reading: recognizing the underlying structure of the text, particularly as related to mathematics word problems or problems in science that require the application of a formula or principle to arrive at an answer. Other, more general situations, reinforce key principles or properties of an important mathematics or scientific concept or idea. Students have ownership of the process as they employ their own creativity in demonstrating mathematics or science content. As students exchange cards, they have additional opportunities to engage in reading a similar problem and demonstrating the application of key concepts or processes emphasized in the lesson.

Implementing the Strategy

1. Students are given instruction on creation of their story card. Instructions should clearly identify the concept or idea that should be the focus of the story problem.

2. Students create their story card addressing the theme or topic.

3. Students also record their solutions in their notebook or other safe keeping place so they can refer to them later.

4. Students exchange cards or the teacher collects and redistributes cards so that each student solves the story problem of another student.

5. Students solve the problem recording their solution and process on the flip side of the card.

6. Cards are returned to the creator who verifies the solution.

7. The teacher responds to any questions or selects cards to use as a review of the concept.

Figure 4.31. Story Card and Solution from Pythagorean Theorem Lesson.

Card 2: Toddle the Turtle starts his day sunning on a nice bright rock. For lunch, he swims due south 7 yards to the local fly-fishing hole. Afterwards, he heads due east for 12 yards to visit Perry the Pelican's humble abode. How far away is Toddle's rock from Perry's abode, as the crow flies? [Record your answer to the nearest tenth of a yard.]

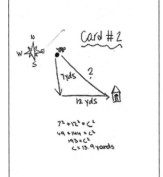

Modifications and Other Considerations

Though this strategy is readily applied to mathematics, it can also be implemented to reinforce science applications. Scientific concepts have many applications. The story card concept can be used to reinforce study of those concepts. The story cards can be creative opportunities for students to explore scientific concepts. Answer cards can include illustrations and diagrams to further demonstrate understanding of the concept. A fourth grade teacher related that she used the story cards in science by having students draw a diagram or picture illustrating a science concept they had studied and read about in their text. When the cards were exchanged, the student receiving the card could use the text to create a story situation or problem that matched the illustration or diagram. For example when studying rocks and their properties, students drew a picture of a certain type of rock, including the label. When students exchanged cards, they were to use their text and notes to identify the properties of that particular rock and write a story situation that emphasized those properties.

Summarizing

Reading Purposes: Retrieving Information
Interpreting
Reflecting and Evaluating

Description

Summarizing is a strategy that requires taking blocks of text and reducing them to essential information. This information typically includes identification of key ideas or main points, essential terminology, and an overview of steps and processes central to the idea or theme developed in the text. Summarizing helps students develop skills in identifying main ideas, using appropriate terms and phrases to describe those ideas, breaking complex ideas or concepts into related parts, developing skills in describing complex phenomena or ideas in concise and precise language, and capturing from text succinct but complete ideas related to a topic. It is important when first using this strategy to model the process with a short piece of text and give students opportunities to work with smaller blocks of text before moving to larger pieces. Some teachers break larger sections of text into smaller pieces and have students summarize the individual sections before writing a synthesis of the multiple summaries.

Implementing the Strategy

1. Identify an appropriate block or passage of text.

2. Students identify key concepts or ideas while also identifying information that contains extra wording.

3. Students summarize these key ideas using appropriate vocabulary and phrases from the reading.

Figure 4.32. Summarizing Grid Used for Percent Lesson.

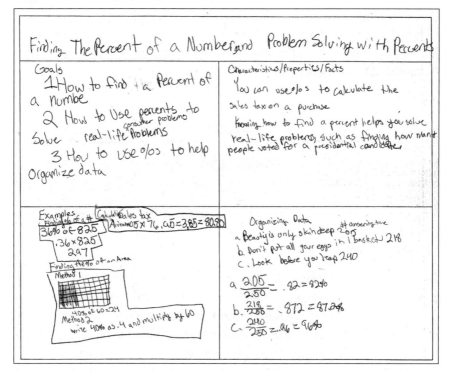

Modifications and Other Considerations

Teachers recognize that teaching students to summarize well is a difficult task. It requires patience, extensive modeling, and lots of feedback; however, the strategy is essential across grade levels and subject areas. Highlighting is a good strategy to develop in conjunction with this particular one. Some teachers also use semantic mapping as a way of helping students identify key concepts and ideas. Implementing several strategies at the same time can be done effectively if the various strategies support a larger goal and if students have the appropriate backgrounds to use the multiple strategies. In the example above, the teacher provides students with goals for their reading based on the lesson. Students use a grid to identify key concepts that support the identified goals.

Teachers also report introducing summarizing in small groups allowing students to produce a group product before applying the strategy on their own. Another possibility is to have students share their summaries in small groups and revise them based on the discussion. It is essential that students revisit the text before modifying their summaries so that they see how the information was embedded in the text and how text features can help in identifying key ideas and information. One teacher shared that as she introduces summarizing to students each year as a strategy that she will provide a short paragraph with six to eight sentences that have key pieces of information missing. She reports this is an effective way to introduce the strategy before students use it on their own. A biology teacher pointed out that she requires students to include a diagram or illustration along with their summary (when appropriate). Examples could be required in mathematics as a way of demonstrating how the ideas in the summary are applied. Sometimes the summary might include information about how to apply a strategy or process to a problem or situation.

Survey, Question, Read, Recite, Review (SQ3R)

Reading Purposes: Retrieving Information
Interpreting
Reflecting and Evaluating

Description

SQ3R is a technique focusing on strengthening reading comprehension developed by Robinson (1941). The strategy is designed to be used before, during, and after reading. Survey, Question, Read, Recite, and Review gives students a framework for thinking about their approach to reading for understanding. The strategy promotes active engagement of readers with text as they monitor their comprehension. The strategy works best with texts that require students to process and understand a lot of information. The strategy requires a more careful approach to reading text, thus more time; however, it is a very efficient use of that time in which students are comprehending what they read. The success of the strategy is evident in the number of disciplines, such as law and nursing, that recommend the strategy. The strategy is somewhat complex and cannot be used independently by students without modeling and practice.

Implementing the Strategy

1. Students **survey (S)** a text passage to get an overview of the contents. This focus typically includes the title, the introductory paragraph, headings, summaries, and examples and illustrations.

2. Students develop a **question (Q)** that focuses on the information in that passage. This might result from turning a heading into a question; however, this is not always the most effective technique and may fail to focus on the major idea(s) of that passage.

3. Students **read (R)** to answer the question formulated in step 2. This gives a purpose for reading.

4. Students **recite (R)** to develop understanding and enhance their ability to remember key ideas. Reciting involves stating ideas and information in one's own words.

5. Students **review (R)** to check for understanding. This involves constructing an answer to the question formulated earlier in the process. Review

Figure 4.33. SQ3R from Tradebook Selection.

Survey, Question, Read, Recite, Review for _Smell and Taste_

Survey the text to get an idea of the contents.	Describe how people smell and taste.
Construct a Question that focuses on the information.	How does the tongue help us taste?
Read the passage paying attention to the answer to your question.	The tongue has bumps called papillae on taste buds. Receptors are sensitive to chemicals in foods. Receptors send messages to the brain.
Recite the answer to your question.	The tongue contains papillae (taste buds) that are sensitive to chemicals in foods.
Review your answer and make any necessary changes.	The tongue is covered with bumps called papillae (taste buds) that are sensitive to chemicals in foods allowing us to sense sweet, sour, salty and bitter taste.

Chapter 3: How we smell and taste.

can include reciting answers to multiple questions posed for one or more passages of text.

Modifications and Other Considerations

This example demonstrates how reading strategies can help students comprehend texts such as tradebooks. The teacher in this case used a series of tradebooks about the senses to reinforce conceptual understanding of science (Silverstein, Silverstein, & Nunn, 2002). The student selected a question of interest based on the reading thus developing a purpose for reading the particular chapter. Notice how read, review and recite results in more concise statements related to the question. Such progression does not occur automatically. Students in this class were familiar with the SQ3R process as evidenced in their facility with the use of the strategy. Students ultimately shared their findings in a larger group.

SQ3R is considered an effective strategy when working with Limited English Proficient students especially when students are given some scaffolding that moves them through the process. Such scaffolding can include graphic organizers with the steps or classroom visuals that remind students of what they should do at each step. Teachers add that classroom discussion to get a sense of students' prior knowledge is also important, but note that many students may not have an extensive background when it comes to covering a complex mathematics or science concept. Discussion used before using SQ3R can be helpful in bridging the gaps in what students bring to the experience. One teacher notes that this discussion is also important when using nonfiction texts such as tradebooks to bring students' prior knowledge to the forefront. Those who use SQ3R report that adding a writing component (Call, 1991). Discussion and writing then emphasize the modes of communication: speaking, reading, listening, and writing.

Another modification emphasizes the importance of attending to visuals in comprehending text. In science and mathematics, it is important that students pay attention to the graphs, illustrations, examples, and figures that frequently are important elements in the text. It is important that surveying include paying attention to these visual components. One teacher has modified the process to emphasize these visual elements by adding Visualize to the process. Survey-Visualize-Question-Read-Recite-Review. This emphasis underscores the important of graphical aids and helps students make connections between the text elements and the concepts reinforced and clarified through visuals.

T-Chart

Reading Purposes: Retrieving Information
Interpreting
Reflecting and Evaluating

Description

T-charts are simple graphical devices that consist of two columns that allow the development of multiple reading goals. T-charts can be used with a wide range of text types including both fiction and nonfiction, as well as an organizational tool for recording ideas during discussion. The strategy helps students organize their thinking and facilitates the understanding of relationships between and among concepts, thus extending students' critical thinking abilities. Used effectively the strategy requires students to examine concepts to identify characteristics, properties, and other elements that are related to concepts described in the text. This simple organizational tool presents students with a purpose for reading and helps focus their attention on details that they may have missed without a means of organizing their thinking while reading.

Implementing the Strategy

1. Identify two major themes, ideas, or concepts from the text.

2. Write one of the words or phrases over each side of the t-chart.

3. Locate information in the reading, or construct responses based on the reading, that fit into each of the columns of the t.

4. Synthesize results through discussion, writing activities, or other method of summarizing the information from the t-chart.

Figure 4.34. T-Chart for Square versus Rectangle.

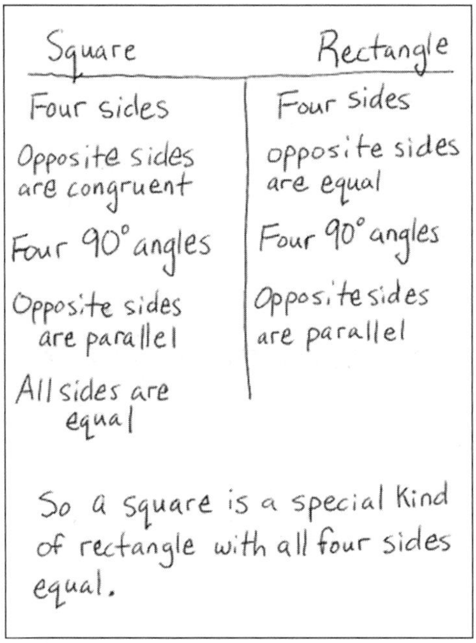

Square	Rectangle
Four sides	Four sides
Opposite sides are congruent	opposite sides are equal
Four 90° angles	Four 90° angles
Opposite sides are parallel	Opposite sides are parallel
All sides are equal	

So a square is a special kind of rectangle with all four sides equal.

Modifications and Other Considerations

In this example, the t-chart effectively helped the student make an important conclusion about squares and rectangles. By comparing characteristics from the text, it becomes clear that rectangles and squares have the same characteristics except squares have all sides being equal. This allows the student to conclude that a square must be a special kind of rectangle.

T-charts can have multiple foci. In science for example, they can help students understand the difference between inferences and observations. Students note the events in the text, record the event in the appropriate column (observation or inference) and then construct, or locate in the text, the observation or inference that corresponds to that statement. T-charts also work for pros/cons, variables such as x/y or x/f(x), similarities/differences, input/output, cause/effect, fact/opinion, and other pairs of concepts or ideas.

Another modification of the t-chart is to make the leg of the T wide enough to include suggestions for categories that could be addressed using the labels on each end of the T. Such as comparing/contrasting mammals/reptiles (labels along top of T) along dimensions such as body temperature, reproduction, diet, habitat, and so on (appear inside the leg of the T). In math, this wider leg can be used to draw examples, show work, or state an appropriate rule or mathematical principle.

One elementary teacher uses t-charts as an easy, yet effective, tool for recording information during oral reading of a text passage. As ideas emerge in the reading the class pauses, identifies the information to fit into the t-chart, and then records the information. This helps students see how concepts are related and helps focus group reading. A middle school teacher shared that he often uses the t-chart as a pre-reading organizational tool; students revise it during the lab or lesson, and then write a summary of the information as part of the night's homework.

Think, Pair, & Share

Reading Purposes: Retrieving Information
Interpreting
Reflecting and Evaluating

Description

This is a cooperative learning strategy that involves students in discussing key ideas or concepts presented in a lesson. The strategy reinforces both individual and shared learning. Think, pair, and share (Lymna, 1981) provides organization that gives structures to students' exploration of key concepts. The think time that is built into the strategy promotes reflection and improves the quality of student responses. Students should be encouraged to draw from the prior experience or background knowledge and to reinforce this thinking with ideas and information from the text (which can also include class notes). The pair and share components provide active learning experiences that improve students' development of critical concepts and conceptual understanding. This strategy is especially appropriate for classrooms new to cooperative learning approaches since it provides a low-risk opportunity for students to work together while making sense of text or classroom information.

Implementing the Strategy

1. Teacher identifies a key question from a lesson or reading and shares the question or problem with students.

2. Students THINK about the question or problem individually. Students should be encouraged to reflect on the question or problem before recording any preliminary thinking.

3. Students pair with a peer to share their ideas and thoughts about the question or problem. Students discuss their perspectives, approaches, and understanding of the problem or concept. The idea is for students to synthesize their ideas, question information, and provide justification for thinking. Teachers should limit the time so that students focus on the task.

4. The teacher calls on pairs of students to share their thinking about the question or problem.

5. Students record any additional ideas or information that emerge from the discussion and/or correct any errors in their logic or thinking.

Figure 4.35. Think-Pair-Share from Lesson on Systems.

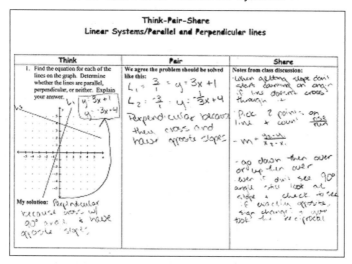

Modifications and Other Considerations

There are several modifications to this popular strategy that can be found in the literature. Listen-Think-Pair-Share is one modification where students are required to listen carefully to the presentation of a problem or other information before responding to a teacher-constructed query. Another modification has student pairs share their information with small groups instead of the entire class. This further promotes group processing and helps students to refine their thinking. Teachers may then ask each of the larger groups to provide a brief summary or may call on one group to give an overview of their conclusions.

One math teacher reported adapting this approach by having students record their initial thoughts on a problem or question using a grid such as the one provided in this text. Students then exchange papers with a peer and THINK again about the problem considering their peer's ideas. Then the process continues with sharing in pairs and then whole class discussion. This extra step gives students an opportunity to consider differences and similarities in thinking before discussions with a peer. A science teacher reported that the THINK part of the activity is sometimes done as part of homework or as a warm-up to get students thinking about a topic before class begins.

Tree Diagram

Reading Purposes: Retrieving Information
Interpreting

Description

Tree diagrams are an effective strategy to help students visualize relationships in decision-making processes. Tree diagrams provide a way of organizing terms or ideas that are part of classifying or sequencing. The diagrams provide thinking tools that help students see how events are related. Trees are a powerful problem solving strategy in mathematics to see how combinations and outcomes of events can be organized. Tree diagrams are also highly effective in showing multiple options that can be made in decision making. This strategy supports students as they evaluate possible outcomes and relate them to properties or principles of mathematics and science. Tree diagrams require students to think about starting points and map out possible outcomes. These outcomes become organized in a way that assists the user in making reasoned conclusions about events depicted in the tree diagram.

Implementing the Strategy

1. Start with one or more events or categories.
2. Draw a line and list a possible outcome or event for each original word or phrase. Draw additional lines and list all other possibilities.
3. Reflect on the possibilities and determine additional outcomes or events that could result from each. Draw lines to possible outcomes and events as necessary.

4. Review the completed tree diagram and use the information to make conclusions about the event or problem that is represented.

Figure 4.36. Tree Diagram Showing Dessert Combinations.

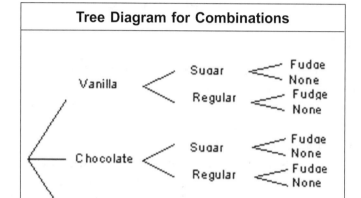

Modifications and Other Considerations

This strategy can be used individually or in small and whole-class settings. The strategy is an effective way to help students understand stages in sorting or classifying information. In science, a primary grades class used the tree diagram to better understand "living versus nonliving things." Branches from living went to plants and animals. Branches from nonliving went to solid, liquid, and gas. Tree diagrams can be modified to fit the appropriate number of outcomes or events. Tree diagrams can also be used to show the relationship between main ideas from text. Teachers may provide a tree diagram with various parts completed to help guide students in making associations between events or outcomes. One teacher reported frequently giving students a completed tree diagram as a guide to direct their reading. After reading, students were engaged in discussing parts of the text that supported the construction of the tree diagram. As such, students were using text to evaluate the accuracy of the tree diagram, thus synthesizing key concepts and information represented in the outcomes and events in the diagram. Students may also present completed tree diagrams as a part of discussing or sharing what they learned about a topic or problem.

Venn Diagram

Reading Purposes: Retrieving Information
                          Interpreting

Description

Venn diagrams have a long history as a strategy that helps students make logical distinctions about the similarities and differences between concepts or ideas. In applying the strategy, students are able to list features, properties, or characteristics of the two concepts in the respective circles, preserving the common overlapping sections to record any shared elements. The strategy provides an organizer than gives students a purpose for their interaction with text or the lesson. Venn diagrams promote critical thinking by helping students consider how properties, features, components, and so on. between two concepts are related. In mathematics, fore example, the diagrams provide powerful tools for promoting skills in reasoning and logic. An Internet search for Venn diagrams will return numerous links with helpful generators.

Implementing the Strategy

1. Identify concepts, items, and so on that are to be compared and contrasted.

2. List characteristics that belong only to the items or concepts being considered in the far right and far left parts of the circles.

3. List characteristics that the two items or concepts have in common in the center, overlapping parts of the circles.

4. Summarize the differences and the similarities between the ideas, concepts, or items.

Figure 4.37. Venn Diagram for GCF and LCM.

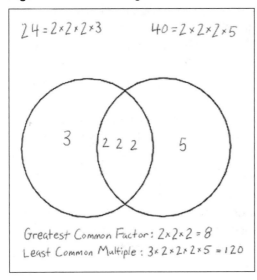

$24 = 2 \times 2 \times 2 \times 3$ $40 = 2 \times 2 \times 2 \times 5$

3   2 2 2   5

Greatest Common Factor: $2 \times 2 \times 2 = 8$
Least Common Multiple : $3 \times 2 \times 2 \times 2 \times 5 = 120$

Modifications and Other Considerations

The illustration is an interesting application of the Venn diagram that illustrates the logic behind the strategy. The overlapping circles are used to show factors that the two numbers (24 and 40) have in common. The two non-overlapping regions show the remaining factors of the two numbers. Greatest common factors are based on 'common' factors of two or more numbers. The Venn diagram helps students develop a mental model for their thinking. The least common multiple can be formed by multiplying all the factors appearing in the diagram. Students who use Venn diagrams in this way are building important logical thinking skills that will help them use this strategy in other situations where they are looking for common and dissimilar characteristics.

Diagrams can also be constructed that allow the consideration of three concepts or ideas. It is recommended that students are proficient using the standard two-circle diagrams before moving to more complex extended diagrams that compare and contrast three ideas, concepts, or items. Students have good general skills comparing things as part of their everyday life. It is important in using Venn diagrams that teachers support careful reflection on important similarities and differences.

Elementary teachers sometimes place two hula-hoops on the floor allowing them to overlap. Initially, it is good to introduce or review the strategy with topics and examples that everyone will understand. Students place cards, pictures, or phrases in the appropriate sections of the diagram. This provides an active way to engage students in using Venn diagrams. Sticky notes may also be used to complete class Venn diagrams on the board or to structure sharing students' responses to a Venn diagram activity. The important element is that students are actively constructing or completing the diagrams, though teachers may initially engage in various levels of scaffolding and modeling to help students become familiar with the strategy.

Visual Association

Reading Purposes: Retrieving Information
                           Interpreting
                           Reflecting and Evaluating

Description

Mathematics and science texts contain many visual elements that represent key concepts, ideas, and vocabulary. These visuals are an integral component of the text and students should develop skills in using these visuals in their comprehension of the material. Visual association is especially effective in helping students master critical vocabulary though the process of recalling images that often depict the complexity of concepts embedded in mathematical and scientific concepts. Forming visual associations is often stressed as a means of improving memory. Visual associations include the use of diagrams, tables, pictures, charts, graphs, models, and so on. This strategy requires students to attend to visuals in the text and interpret how the visual relates to key concepts. Students respond to a visual by describing how the visual is associated to the main ideas or concepts in the reading and extend their thinking by constructing or describing a similar visual which cements their association of the concept or idea with the visual.

Implementing the Strategy

1. The teacher identifies one or more visual(s) that are important to understanding the concept presented in the text.

2. Students sketch the visual and describe how the visual relates to the main idea or concept of the reading.

3. Students describe or construct a similar visual that supports the same idea or concept as the original.

4. Students write two to three sentences relating the original and similar visuals to each other and the main idea or concept from the reading.

Figure 4.38. Visual Association for Slopes of Lines Lesson.

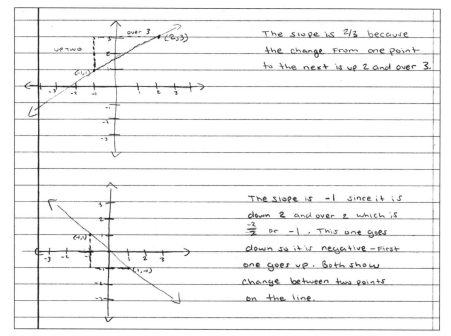

Modifications and Other Considerations

Teachers initially identify one or more visual components that he or she feels is essential to promoting students' understanding of the text. As students become familiar with the strategy and are successful in using it, they can take more responsibility for selecting the visuals that help support their learning of the concept. This is not an easy transition to make, particularly since most students may not have been taught to process these visual elements as part of their reading. When moving to independence, it is important for teachers to have students share their selections, descriptions, and similar visuals along with some discussion about why that particular visual was central to understanding a key concept or idea. This helps students see the way that others think about using the visuals to support their learning.

Students who do not have strong visualization skills may need additional support in creating visual images and associating them to important ideas and concepts. Teachers may provide hints and ask questions to help students think of similar visuals. One teacher noted that sometimes it is necessary to modify the strategy so students are not required to generate another visual that compares to the original. She stressed, however, that the class would discuss the visual component and their descriptions and then together would come up with another way of representing the concept or idea. She found that this was successful with complex ideas and concepts, particularly models that had several steps or components.

Vocabulary through Context

Reading Purposes: Retrieving Information
Interpreting

Description

This strategy emphasizes the importance of context in building an understanding of vocabulary. It is important for students to build skills in determining the meaning of new words based on context clues. Though students may have experiences using context clues for unfamiliar words, they may not be aware of the usefulness of this approach in mathematics and science. This strategy helps students attend to definitions, explanations, examples, and elaborative details to determine the meaning of words. Students write the sentence from their text where the word was used. They follow up by writing a definition and then a student-constructed sentence using the word. To further connect the word to the concepts, students draw a picture or illustration or connect the word with something they are familiar such as a song, a story, or a definition, explanation, restatement, example, and punctuation.

Implementing the Strategy

1. Before students read the material, the teacher lists vocabulary words related to a passage the students will read.

2. As students encounter each word, they pause and write the sentence in which the word appears.

3. Have students try to deduce the word's meaning by context clues and make some notes about the expected meaning.

4. Students write a definition for the word using a dictionary, glossary, or other source.

5. Students write the word in a sentence that demonstrates they understand the meaning of the word.

6. Students create a visual link to the word by thinking of a picture, song, experience, or other tool that will help them remember the meaning of the word.

7. Students may share their strategies and ideas after the reading.

Figure 4.39. Vocabulary through Context Example for Right Triangles.

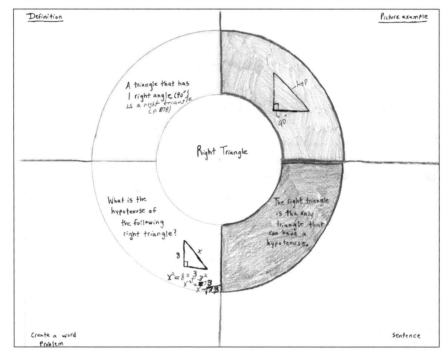

Modifications and Other Considerations

This strategy can be used in paired readings giving students an opportunity to work together in using the strategy to facilitate understanding of unfamiliar vocabulary. As students become more competent in using the strategy, they can begin to identify words to focus on instead of depending on the teacher to provide a list. Teachers can have students share their selections as part of the lesson development to guarantee that students are focusing on words that may pose a challenge to students.

Some teachers require students to apply the strategy to at least one additional term that they encounter in their reading that is unknown or unclear to them or a word with which they are familiar but need to review. This additional step helps students take ownership of the strategy while helping them reflect on how well they understand how terms are used in their reading.

Often students skim over words they have seen before but do not consciously consider what the word means. Understanding the meaning of a word is not the same as recognizing a word.

Word Sort

Reading Purposes: Retrieving Information

Description

This vocabulary development strategy requires students to sort terms or phrases into categories. One goal is to help them see the relationship between words and important concepts that they encounter in their reading. During the activity, students sort words written on cards or the board based on similarities, differences, relationships, or other criteria. More complex word sorts are appropriate as post-reading activities while a word sort with basic words that students will encounter in the lesson can be used to activate prior knowledge or as a pre-assessment activity to determine students' current familiarity with relevant words and concepts. Students in primary grades often engage in word sorts as they learn word sounds such as sorting by how prefixes, suffixes, or root words are spelled. Word sorts in mathematics and science reinforce concept groups while also building word recognition and vocabulary skills.

Implementing the Strategy

1. The teacher identifies two or three classification words or phrases (categories) and a list of words that are associated with each label. The words or phrases should come from the students' readings.

2. Students receive the words written on index cards or other blocks of paper. The cards or slips of paper should be shuffled so that they are not in order. All of the words can also be written on a sheet of paper with the categories clearly identified. For example the word to sort could appear in random order listed in several columns. A table with several columns could be provided for students to sort the words or phrases.

3. The teacher gives the students the classification labels. If these words or phrases are included on the cards, students can locate them and use them to sort their stacks.

4. Students work individually, in pairs, or in small groups sorting the words into their appropriate category.

5. Students share their outcomes and the teacher uses discussion to correct any errors in sorting the cards. It is important for the teacher to monitor student progress during the activity so that students with different sorts can discuss the thinking that led to their classification.

6. Students should be allowed to "resort" cards based on class discussion and feedback.

Table 4.5.   Word Sort for Fourth Grade Unit on Biomes of North America.

Teacher Determined Categories	Student Words to Sort	
Deciduous Forests	Great Plains	Distinct Seasons
Grassland	Eastern United States	Drought Resistant Shrubs
Desert	Cactus	Trees Lose Leaves
Chaparral	Snowshoe Hare	Coastal California
Boreal Forest	Small Waxy Leaves	Mediterranean Climate
	Nocturnal Mammals	Coniferous Forests
	Moderate Climate	Primary Biome of Canada
	Sandy Soil	Prairie
	Grasses, Flowers, Herbs	Limited Rainfall
	Extreme Daily Temperatures	Southwestern United States

Modifications and Other Considerations

Word sorts can be closed or open. In closed sorts, the teacher provides the categories for the students. In an open sort, students come up with the categories that they use for sorting the words. Students should be encouraged to provide a justification for their classification system and should ground their thinking in their studies of mathematics or science. It is recommended that teachers using open word sort model using several sets so that students can see how the teacher's thinking assists in classifying the words or phrases into groups.

In primary grades, pictures may replace words. For example, students might place pictures related to the four seasons in appropriate categories. Sorting of pictures (diagrams) is also a powerful geometry strategy. Students could sort geometric figures based on their characteristics (such as regular vs. nonregular polygons).

This strategy can be extended to include processes that involve interpreting information by requiring students to compare and contrast the terms or phrases as part of their justification for their classification system. One teacher modifies this strategy by writing terms on sticky notes which she distributes to students. Students have to decide in which category their term belongs. They come to the board and place their term giving a rationale for their decision. Once all the sticky notes are placed, the teacher engages students in discussing the categories and the terms that appear in each. Through discussion, words may be relocated but only after the class agrees that the placement is the best choice.

Word Walls

Reading Purposes: Retrieving Information

Description

Word walls provide an opportunity to emphasize key vocabulary words that are a central part of study. Word walls can take on many formats from simple display of words, sometimes with associated definitions, to more elaborate presentations that extend beyond words and their definitions to include key concepts, illustrations, and other pertinent information. Some teachers use word walls as an interactive teaching tool requiring student contributions as new words and their related information are displayed as they are encountered in the lessons. Word walls provide students with reminders of terminology that they should use in their communication about mathematical and scientific concepts and processes. Word walls can become cumulative displays of vocabulary over an entire year or they can be used to reinforce vocabulary for a specific unit of study. Regardless of the time frame or elaboration of display, word walls provide a ready source of information for students.

Implementing the Strategy

1. Add words as they become pertinent to the lesson or reading.

2. Display the words in a prominent place so they are easily visible by students. Use cards or sturdy paper so the words can remain on display for an extended period of time.

3. Consider using colored backgrounds or groupings to show associations between words.

4. Students should review words as they are added by providing opportunities for students to identify the word(s) in their reading, as well as using the word(s) in their writing and other forms of communication.

Figure 4.40. Student-Generated Word Wall Cards for Benchmark Fractions.

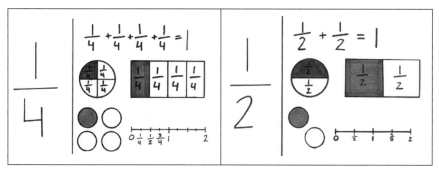

Modifications and Other Considerations

Word walls could be considered as one application of a word sort. Teachers frequently display words in dictionary format that includes parts of speech and definitions. Some teachers add a sentence showing the word being used in the context likely to be encountered in students' study or reading. Math teachers frequently add a diagram, illustration, or example showing the concept behind the word.

Many teachers make word walls interactive. The development of word walls should be ongoing. One teacher requires students to keep organized index card collections (in a file box) of words from the word wall. As the word is introduced and displayed, students write the word on an index card along with the definition. As students read and encounter the word, they flip the card over and rewrite one sentence where the word appeared. Students are encouraged to write the sentence in their own words. Another teacher has students work in small groups to create word wall entries. Four to five words are identified and each group constructs an entry for the wall. The groups present their word, the definition, and discuss how the word is used in their reading as they add their entry to the word wall. One science teacher displays words in groups that have a particular relationship or are associated to a particular concept. When a group of words is complete, the students work in small groups to write a short paragraph that uses all the words. Another teacher who has several ESL students, finds the word in the students' languages and adds those to the bottom of the word wall card. This gives ESL students some support as they learn new and often technical vocabulary. Students who know the word in their native language are more likely to use the word in communicating with parents about their learning. These various interactive methods extend the use of word walls by making them a central part of learning.

Writing to Develop Understanding

Reading Purposes: Retrieving Information
Interpreting
Reflecting and Evaluating

Description

This strategy can be used in conjunction with any reading strategies that help students organize information from their reading. While writing should be an integral part of a solid reading program, this strategy provides a structure for students supporting their understanding of key concepts from the reading. Teachers may focus on specific information by breaking larger text passages into smaller chunks. This strategy is most helpful if the text passage focuses on a main idea or concept. The strategy engages students in following key steps in identifying information from the text. As students read and identify the necessary information, they are organizing their thinking and supporting their understanding from the text. Through the strategy, students have a foundation that will allow them to extend their work by completing a piece of writing based on their work.

Implementing the Strategy

1. Students identify the main idea from their reading and record this information, preferably in sentence form.

2. Students identify a major detail to explain or support the main idea.

3. Students provide minor details that describe or explain the major detail.

4. Students write a summary statement.

5. Students complete a short piece of writing based on their work from steps 1 through 4.

Figure 4.41. Writing to Develop Understanding on Factors Lesson.

Main Idea

A factor is a number multiplied to get a product.

Major Details

Factors are prime or composit.
Prime numbers only have 1 and the number as factors. Composit numbers have more than two factors.

Minor Details/Supporting Information

Composit numbers can be broken into prime factors.

$12 = 2 \times 6 = 2 \times 2 \times 3$

Summary

A factor tree helps break a number into prime factors.

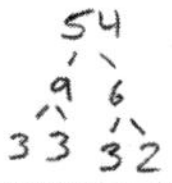

54
9 6   or $3^3 \times 2$
3 3 3 2

Paragraph

Factors can be broken down into two categories. They can either be composit or prime numbers. Composit numbers can be broken into prime factors while prime numbers can only be broken down into one and the number itself. You can use a factor tree to break down a composit number into prime factors.

Modifications and Other Considerations

This strategy helps students read and write with a clear purpose. The strategy supports students' thinking about key concepts. Students sometimes experience various levels of difficulty with this strategy. Teachers frequently use this strategy several times in whole class or group work before students are expected to use it on their own. Once students are comfortable with the strategy, they can use it on their own. It is important, particularly when students are working on their own, that there are opportunities for some sharing of the work. Teachers might have one or two students share their work and then engage the class in a discussion based on the information. Some teachers revisit the writing after they complete some additional instruction allowing students an opportunity to refine any ideas before it becomes a final piece of writing. The products become an important resource in reviewing for assessments.

One teacher uses this strategy with each new section in a chapter. She has students keep their work in a notebook and assesses the work at the end of the unit as part of a class participation grade. Another teacher asks students to include their best writing as part of a portfolio of their work for the grading period or unit. One middle grades teacher shared that he allowed students to use these products during the last five minutes of a test. This provided motivation for students in completing the initial assignment. He found that it also helped students be more reflective during tests because they had to be more aware of which question was difficult for them. He reported that the five minutes was generally only enough time for a student to review their writing and change one test item.

Table 4.6.   Summary Table of Strategies with Reading Purpose

Reading Strategy	Retrieving Information	Interpreting	Reflecting and Evaluating
Anticipation Guide	X	X	X
Brainstorming	X		
Cause and Effect	X	X	X
Cloze Vocabulary Exercise	X		
Compare & Contrast Summary	X	X	X
Concept Circle	X	X	X

Table 4.6. Summary Table of Strategies with Reading Purpose
(Continued)

Reading Strategy	Retrieving Information	Interpreting	Reflecting and Evaluating
Concept Synthesis	X	X	X
Cubing	X	X	X
Cue Cards	X		
Fact Summary	X	X	
Five Minute Pause	X	X	X
Frayer Model	X	X	X
Highlighting	X	X	
Inquiry Chart	X	X	X
Jigsaw	X	X	X
Journaling	X	X	X
K-W-L	X	X	X
Learning Log	X	X	X
List-Classify-Identify	X	X	
Parallel Notes	X	X	X
Prefix Mastery	X		
Presentation Project	X	X	X
Problem Solving Seven	X	X	X
Question Answer Response (QAR)	X	X	X
Role, Audience, Format Topic (RAFT)	X	X	X
Reciprocal Teaching	X	X	X
Semantic Feature Analysis	X	X	X
Semantic Mapping	X	X	
Sentence Cards	X	X	
Sequence Forms	X	X	X
Story Card	X	X	X
Summarizing	X	X	X
Survey, Question, Read, Recite, Review (SQ3R)	X	X	X
T-Chart	X	X	X
Think Pair Share	X	X	X
Tree Diagram	X	X	
Venn Diagram	X	X	
Visual Association	X	X	X

Table 4.6.   Summary Table of Strategies with Reading Purpose
(Continued)

Reading Strategy	Retrieving Information	Interpreting	Reflecting and Evaluating
Vocabulary Through Context	X	X	
Word Sort	X		
Word Walls	X		
Writing to Develop Understanding	X	X	X

References

Aarnoutse, C. & Schellings, G. (2003). Learning reading strategies by triggering reading motivation. *Educational Studies, 29#4*, 387–409.

Adams, T.L. (2003). Reading mathematics: More than words can say. *The Reading Teacher, 56#8*, 786–795.

Allington, R. L. (2002). You can't learn much from books you can't read. *Educational Leadership, 60*(3), 16–19.

Ansberry, K. B. & Morgan, E. (2005). *Picture Perfect Science Lessons: Using Children's Books to Guide Inquiry*. Arlington, VA: National Science Teachers Association Press.

Armbruster, B. B., Anderson, T. H., & Ostertag, J. (1989). Teaching text structure to improve reading and writing. *The Reading Teacher, 43*(2), 130–137.

Armbruster, B. B., Echoles, C., & Brown, A. (1983). *The Role of Metacognition in Reading to Learn: A Developmental Perspective*. Urbana, IL: Center for the Study of Reading.

Aronson, E., Blaney, N., Stephin, C., Sikes, J., & Snapp, M. *The jigsaw classroom*. (1978). Beverly Hills, CA: Sage Publishing Company.

Aulls, M. W. (2003). The influence of a reading and writing curriculum on transfer learning across subjects and grades. *Reading Psychology, 24*, 177–215.

Balfanz, R. & MacIver, D. (2000). Transforming high poverty urban middle schools into strong learning institutions: Lessons from the first five years of the talent development school. *Journal of Education for Students Placed at Risk, 5#1/2*, 137–158.

Ball, D. L. & Bass, H. (2003). Toward a practice-based theory of mathematical knowledge for teaching. In B. Davis & E. Simmt (Eds.), *Proceedings of the 2002 Annual Meeting of the Canadian Mathematics Education Study Group*, pp. 3–14. Edmonton: Canadian Mathematics Education Study Group.

Barton, M. L., Heidema, C. & Jordan, D. Teaching reading in mathematics and science. *Educational Leadership, 60*(3), 24–28.

Berghoff, B. (1993). Moving towards aesthetic literacy in the first grade. In D.J. Leu & C. Kinzer (Eds.), *Examining central issues in literacy research, theory, and practice: Forty-second yearbook of the National Reading Conference* (pp. 217–226) Rochester, NY: National Reading Conference.

Berghoff, B. (1998). Multiple sign systems and reading. *The Reading Teacher, 51#6*, 520–523.

Borasi, R. & Siegel, M. (2000). *Reading Counts.* New York: Teachers College Press.

Brown, G. T. L. (2003). Searching informational texts: Text and task characteristics that affect performance. *Reading Online, 7*(2). [Available online at http://www.readingonline.org/articles/.]

Bryant, D. P., Ugel, N., & Hamff, A. (1999). Instructional strategies for content-area reading instruction. *Intervention in School and Clinic, 34*(5), 293–304.

Budiansky, S. (2001, February). The Trouble with Textbooks. *Prism.* [Available online at http://www.prism-magazine.org/feb01/html/textbooks.cfm.]

Call, P. E. (1991). SQ3R + what I know sheet = one strong strategy. *Journal of Reading, 35,* 50–52.

Chi, M.T.H. (2000). Cognitive understanding levels. In. A. E. Kazkin (Ed.), *Encyclopedia of Psychology*, 2:146–151. London: Oxford University Press.

Collins, N. D. (1994). *Metacognition and Reading to Learn.* Bloomington, IN: ERIC Clearinghouse on Reading, English and Communication. [ED 276427]

Cook, Doris M. (1989). *Strategic Learning in the Content Areas.* Milwaukee: Wisconsin Department of Public Instruction.

Cotton, K. (1988). Classroom questioning. *School Improvement Research Series.* Portland, OR: Northwest Regional Educational Laboratory. [Available online at http://www.nwrel.org/scpd/sirs/3/cu5.html.]

Dickinson, D. K. & DiGisi, L. L. (1998). The many rewards of a literacy-rich classroom. *Educational Leadership, 55#6*, 23–26.

Diezmann, C. M. & English, L. D. (2001). Promoting the use of diagrams as tools for thinking. In A. A. Cuoco & F. R. Cuoco (Eds.), *The Role of Representation in School Mathematics*, pp. 77–89. Reston, VA: National Council of Teachers of Mathematics.

DiGisi, L.L. & Yore, L.D. (1992). *Reading comprehension and metacognition in science: Status, potential and future direction.* Paper presented at the Annual Meeting of the National Association for Research in Science Teaching, Boston, MA.

Douville, P. & Pugalee, D. K. (2003). Investigating the relationship between mental imaging and mathematical problem solving. *The Mathematics into the 21st Century Project: Proceedings of the International Conference.* Brno, Czech Republic. [Available at: http://math.unipa.it/~grim/21_project/21_brno_03.htm].

Draper, R. J. (2002). School mathematics reform, constructivism, and literacy: A case for literacy instruction in the reform-oriented math classroom. *Journal of Adolescent and Adult Literacy, 45#6,* 520–529.

English, L. (1999). Reasoning by analogy: A fundamental process in children's mathematical learning. In L. V. Stiff & F. R. Curcio (Eds.), *Developing Mathematical Reasoning in Grades K–12,* pp. 22–36. Reston, VA: National Council of Teachers of Mathematics.

Espin, C. A., & Deno, S. L., Performance in reading from content area text as an indicator of achievement feature article. *Remedial and Special Education, 14*(6), 47–59.

Fisher, K. Wandersee, J. H., & Moody, D. E. (2000). Mapping Biology Knowledge. Dordrecht, The Netherlands: Kluwer Academic Publishers.

Fleisher, P. (2002). *Objects in Motion: Principles of Classical Mechanics.* Minneapolis, MN: Lerner Publishing Group.

Flick, L. B., & Lederman, N. G. (2002). The value of teaching reading in the context of science and mathematics. *School Science and Mathematics, 102#3,* 105–106.

Frayer, D., Frederick, W. C., and Klausmeier, H. J. (1969). *A Schema for Testing the Level of Cognitive Mastery.* Madison, WI: Wisconsin Center for Education Research.

Freitag, M. (1997). Reading and writing in the mathematics classroom. *The Mathematics Educator, 8#1,* 16–21.

Fuentes, P. (1998). Reading comprehension in mathematics. *Clearing House, 72#2,* 81–88.

Garner, J. K. & Bochna, C. R. (2004). Transfer of a listening comprehension strategy to independent reading in first grade students. *Early Childhood Education Journal, 32*(2), 69–74.

Gee, T. C. & Rakow, S. J. (1990). Helping students learn by reading: What experienced social studies teachers have learned? *Social Education, 54#6,* 398–401.

Graves, M. F., Juel, C., & Graves, B. B. (2004). *Teaching reading in the 21ˢᵗ century*. Needham Heights, MA: Allyn & Bacon.

Grossen, B. & Romance, N. R. (1994). Science: Educational tools for diverse learners. *School Psychology Review, 23*(3), 442–463.

Guthrie, J.T. (2001, March). Contexts for engagement and motivation in reading. *Reading Online, 4*(8). Available: http://www.readingonline.org/articles/art_index.asp?HREF=/articles/handbook/guthrie/index.html.

Guthrie, J. T., & Cox, K. E. (2001). Classroom conditions for motivation and engagement in reading. *Educational Psychology Review, 13*#3, 283–302.

Guthrie, J. T. & Wigfield, A. (1999). How motivation fits into a science of reading. *Scientific Studies of Reading, 3*#3, 199–205.

Hall, K., Myers, J., & Bowman, H. (1999). Tasks, texts and contexts: A study of reading and metacognition in English and Irish primary classrooms. *Educational Studies, 25*(3), 311–325.

Halliday, M. A. K. & Martin, J. R. (1993). *Writing Science: Literacy and Discursive Power.* Pittsburgh, PA: University of Pittsburgh Press.

Head, M.H. & Readence, J.E. (1986). Anticipation guides: Meaning through prediction. In E.K. Dishner, T.W. Bean, J.E. Readence & D.W. Moore, (Eds.) *Reading in the Content Areas*, 227–233. Dubuque, IA: Kendall-Hunt.

Harnisch, D. L., Polzin, J. R., Brunsting, J., Camasta S., Pfister, H., Mueller, B., Frees, K., Gabric, K., & Shope, R. J. (2005). Using visualization to make connections between math and science in high school classrooms. Paper presented at the annual meeting of the Society for Information Technology and Teacher Education, Phoenix, AZ.

International Reading Association. (2000). *Making a Difference Means Making It Different: Honoring Children's Rights to Excellent Reading Instruction.* Newark, DE: author.

Irvin, J. L. (1997). Building sound literacy learning programs for young adolescents. *Middle School Journal, 28*#3, 4–9.

Jacobs, J. E. & Paris, S. G. (1987). Children's metacognition about reading issues in definition, measurement, and instruction. *Educational Psychologist, 22*#3–4, 235–278.

Jacobson, R. (1998). Teachers improving learning using metacognition with self-monitoring learning strategies. *Education, 118*#4, 579–590.

Jones, B.F., Palincsar, A.S., Ogle, D.S., & Carr, E.G. (1987). *Strategic teaching and learning: Cognitive instruction in the content areas.* Alexandria, VA: Association for Supervision and Curriculum Development.

Kirsch, I., deJong, J., Lafontaine, D., McQueen, J., Mendelovits, J., & Monseur, C. (2002). *Reading for change: Performance and engagement across countries—Results from PISA 2000.* Paris, France: Organization for Economic Co-operation and Development.

Kruidenier, J. (2002). *Research-based principles for adult basic education reading instruction.* Washington, DC: National Institute for Literacy.

Loranger, A. L. (1999). The challenge of content area literacy: A middle school case study. *The Clearing House, 72*#4, 239–243.

Luke, A., & Freebody, P.(1999, August). Further notes on the four resources model. *Reading Online.* Available: www.readingonline.org/past/past_index.asp?HREF=/research/lukefreebody.html.

MacGregor, M. & Price, E. (1999). An exploration of aspects of language proficiency and algebra learning. *Journal of Research in Mathematics Education, 30,* 449–467.

Mathison, C. (1989). Activating student interest in content area reading. *Journal of Reading, 33*#3, 170–176.

Maitland, L. E. (2000). Ideas in practice: Self-regulation and metacognition in the reading lab. *Journal of Developmental Education, 24*#2, 26–36.

McDougal Littell (2005). *Science, Grade 6.* Evanston, IL: Author.

McTighe, J., Seif, E., & Wiggins, G. (2004). You can teach for meaning. *Educational Leadership, 621*(1), 25–30.

Mizelle, N. B. (1997). Enhancing young adolescents' motivation for literacy learning. *Middle School Journal, 28*#3, 16–25.

Moats, L. (1999). *Teaching reading is rocket science: What expert teachers of reading should know and be able to do.* Washington, DC: American Federation of Teachers.

NASA CONNECT. (2002). *Rocket to the Stars.* [Available at http://connect.larc.nasa.gov/.]

National Center for Education Statistics. (2004). *The nation's report card: Reading highlights 2003.* Washington, DC: US Department of Education.

National Center for Education Statistics (2001). *National Assessment of Educational Progress.* Washington, DC: Author. [Accessed online at http://nces.ed.gov/nationsreportcard/.

National Research Council (1995). *National Science Education Standards.* Washington, DC: National Academy Press.

No Child Left Behind Act of 2001, Pub. L. No. 107–110 (2002). [Available at: http://www.ed.gov/policy/elsec/leg/esea02/index.html.]

Novak, J. D., Mintzes, J. J., & Wandersee, J. H. (2000). Epilogue: On ways of assessing science understanding. In J. J. Mintzes, J. H. Wandersee & J. D. Novak (Eds.), *Assessing Science Understanding,* pp. 355–374. San Diego: Academic Press.

Ontario Ministry of Education. (2003). *Early Reading Strategy: The Report of the Expert Panel on Early Reading in Ontario.* Toronto, Ontario: Queens Printer of Ontario.

Organization for Economic Co-operation and Development. (2003). *The PISA 2003 Assessment Framework—Mathematics, Reading, Science and Problem Solving Knowledge and Skills.* Paris, France: Author.

Pape, S. J. & Smith, C., (2002). Self-regulating mathematics skills. *Theory into Practice, 41#2,* 93–101.

Palincsar, A.S. (1986). *Reciprocal teaching: Teaching reading as thinking.* Oak Brook, IL: North Central Regional Educational Laboratory.

Penney, K., Norris, S. P., Phillips, L. M., & Clark, G. (2003). The anatomy of junior high school science textbooks: An analysis of textual characteristics and a comparison to media reports of science. *Canadian Journal of Science, Mathematics and Technology Education, 3*(4), 416–436.

Polya, G. (1957). *How to solve it.* Garden City, NY: Doubleday and Co., Inc.

Pugalee, D. K., Brailsford, I. & Perez, T. (in press). Strategies to support mathematical understanding of LEP students. *Centroid.*

Pugalee, D. K. (2005). *Writing to Develop Mathematical Understanding.* Norwood, MA: Christopher Gordon Publishers, Inc.

Pugalee, D. K., DiBiase, W. J. & Wood, K. D. (1999). Writing and the development of problem solving in mathematics and science. *Middle School Journal, 30*(5), 48–52.

Pressley, M. (1998). *Reading Instruction that Works. The Case for Balanced Teaching.* New York: The Guilford Press.

Robinson, F. (1941). *Effective study.* New York, NY: Harper & Row.

Roeschl-Heils, A., Schneider, W., & van Kraayenoord, C. E. (2003). Reading, metacognition and motivation: A follow-up study of German students in grades 7 and 8. *European Journal of Psychology of Education, 22#1,* 75–86.

Roseman, J. E., Kulm, G., & Shuttlewoth, S. (2001). Putting textbooks to the test. *ENC Focus, 8*(3), 56–59.

Ruddell, M. R. (2005). *Teaching Content Reading and Writing.* New York: John Wiley & Sons, Inc.

Ruddell, R. B. & Ruddell, M. R. (1995). *Teaching Children to Read and Write: Becoming and Influential Teacher.* Boston: Allyn & Bacon.

Schumm, J. S., Vaughn, S. & Saumell, L. (1994). Assisting students with difficult textbooks: Teacher perception and practices. *Reading Research and Instruction, 34*(1), 39–56.

Silverstein, A., Silverstein, V. & Nunn, L. (2002). *Smelling and Tasting.* Brookfield, CT: Twenty-First Century Books.

Stephens, E. C. & Brown, J. E. (2005). *A Handbook of Content Literacy Strategies: 125 Practical Reading and Writing Ideas.* Norwood, MA: Christopher Gordon Publishers.

Taylor, B. M., Pearson, P. D., Peterson, D. S., & Rodriguez, M. C. (2003). Reading growth in high poverty classrooms: The influence of teacher practices that encourage cognitive engagement in literacy learning. *The Elementary School Journal, 104#1,* 3–28.

Thompson, D. R. & Rubenstein, R. N. (2000). Learning mathematics vocabulary: Potential pitfalls and instructional strategies. *Mathematics Teacher, 93*(7), 568–574.

Tovani, C. (2004). *Do I Really Have to Teach Reading?* Portland, ME: Stenhouse Publishers.

Turbill, J. (2002, February). The four ages of reading philosophy and pedagogy: A framework for examining theory and practice. *Reading Online, 5*(6). Available:http://www.readingonline.org/international/inter_index.asp?HREF=turbill4/index.html.

Valverde, G. A. & Schmidt, W. H., (1997/98). Refocusing US math and science education. *Issues in Science & Technology, 14*(2), 60–67.

Wade, S.E., & Moje, E.B. (2001, November). The role of texts in classroom learning: Beginning an online dialogue. *Reading Online, 5*(4). Available online at: http://www.readingoline.org/articles/art_index.asp?HREF=/articles/handbook/wade/index.html .

Wakefield, D.V. (2000). Math as a second language. *The Educational Forum, 64,* 272–279.

Wakefield, A.P. (1998). Support math thinking. *Education Digest, 63#5,* 59–64.

Wandersee, J. H. (2000). Designing an image-based biology text. In J. J. Mintzes, J. H. Wandersee & J. D. Novak (Eds.), *Assessing Science Understanding,* pp. 129–143. San Diego: Academic Press.

Webb, N. (2002). *Depth-of-Knowledge Levels for Four Content Areas.* Madison, WI: National Institute for Science Education and Council of Chief State School Officers.

Weis, I., Pasley, J., Smith P., Banilower, E., & Heck, D. (2003). *Looking Inside the Classroom: A Study of K–12 Mathematics and Science Education in the U.S.* Chapel Hill, NC: Horizon Research Inc.

Wellington, J. & Osborne, J. (2001). *Language and Literacy in Science Education.* Buckingham, UK: Open University Press.

Wiggins, G. & McTighe, J. (2001). *Understanding by Design.* Upper Saddle River, NJ: Prentice-Hall.

Winstead, L. (2004). Increasing academic motivation and cognition in reading, writing, and mathematics: Meaning-making strategies. *Educational Research Quarterly, 28#2*, 30–49.

Woodward, A., & Elliot, D. (1990). Textbook use and teacher professionalism. In D. Elliot & A. Woodward (Eds.), *Textbooks and schooling in the U.S.: 89th Yearbook of the National Society for the Study of Education, Part I*, pp. 178–193. Chicago: National Society for the Study of Education.

Underwood, T. (1997). On knowing what you know: Metacognition and the act of reading. *The Clearing House, 71#2*, 77–80.

Index

About the Author

David K. Pugalee is Associate Professor of Education at the University of North Carolina Charlotte where he is coordinator of the Ph.D. program in Curriculum and Instruction. He earned a Ph.D. in mathematics education from the University of North Carolina at Chapel Hill. He has a bachelor's degree in psychology and master's degrees in mathematics and curriculum supervision. His teaching experience includes teaching mathematics at the elementary, middle, and secondary levels before moving into higher education. He has an extensive list of publications including research articles in *Educational Studies in Mathematics* and *School Science and Mathematics*. His work also includes several books, including being lead author for several in the middle grades *Navigations* series published by the National Council of Teachers of Mathematics. His research interest is mathematical literacy—the relation between language and mathematics learning. He is author of *Writing to Develop Mathematical Understanding*, also published by Christopher-Gordon Publishers, Inc.

W9-BJO-888

The College Panda

SAT Essay

The Battle-tested Guide

ISBN: 978-0-9894964-6-9

*SAT is a registered trademark of the College Board, which does not endorse this product.

For more information, visit thecollegepanda.com

Discounts available for teachers and companies. Please contact thecollegepanda@gmail.com for details.

Table of Contents

Introduction

In the past, the SAT required you to write a persuasive essay on a philosophical question such as *Do rules and limitations contribute to a person's happiness?*

The best approach was well-established: write an example-based response, one based on current events, literature, and history to support your stance. In fact, the top scorers often came into the essay with a slew of examples they had prepared beforehand and could tweak to almost any prompt.

This trend towards "scripted" essays alarmed not only The College Board but also the SAT's critics. How can a standardized test accurately assess the writing abilities of students who are regurgitating memorized sentences in 25 minutes? What's even scarier is that these essays were the ones that did the best.

In light of this, The College Board decided to overhaul the essay in its 2016 redesign of the SAT.

They wanted an essay assignment that would prevent pre-planning and scripted responses. In the face of stiff competition from the ACT, the SAT needed to shed its reputation as a test that could be "gamed."

What they came up with is an analytical essay assignment, one in which you're asked to read a passage and discuss how the author persuades his or her readers. Here's what a typical assignment looks like:

> As you read the passage below, consider how Anthony Simon uses
> - evidence, such as facts or examples, to support claims.
> - reasoning to develop ideas and to connect claims and evidence.
> - stylistic or persuasive elements, such as word choice or appeals to emotion, to add power to the ideas expressed.

> Write an essay in which you explain how Anthony Simon builds an argument to persuade his audience that student competitions should promoted within schools. In your essay, analyze how Simon uses one or more of the features listed in the box above (or features of your choice) to strengthen the logic and persuasiveness of his argument. Be sure that your analysis focuses on the most relevant features of the passage.
>
> Your essay should not explain whether you agree with Simon's claims, but rather explain how Simon builds an argument to persuade his audience.

In theory, students wouldn't be able to regurgitate responses to this type of an assignment.

And even if they could, The College Board would be able to maintain plausible deniability. "But it has a passage that always changes!" they would say.

"There's no way that it's susceptible to prepared examples like it was before."

"So take this test instead of the ACT!"

Yes, prepared examples are no longer relevant to the new format, but it's just as susceptible to planned responses, and I took the SAT myself to prove it, getting a perfect 1600 on the test and a 7/7/7 on the essay.

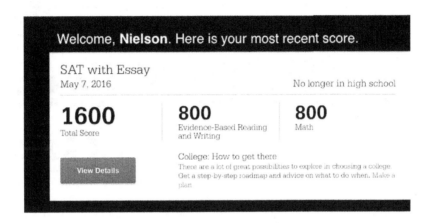

Welcome, **Nielson**. Here is your most recent score.

SAT with Essay
May 7, 2016   No longer in high school

1600
Total Score

800
Evidence-Based Reading
and Writing

800
Math

View Details

College: How to get there
There are a lot of great possibilities to explore in choosing a college.
Get a step-by-step roadmap and advice on what to do when. Make a
plan.

Essay Scores

Essay - Reading	Essay - Analysis	Essay - Writing
7 2 to 8	7 2 to 8	7 2 to 8

Essay Details

In taking the SAT as a teacher, the goal wasn't to get a high score for bragging rights. Nor was it to show you fancy essays that only I could write. **The goal was to develop a framework for a high-scoring essay that could be replicated by all students and applied to all prompts.**

In the process, what I discovered is what I expected all along: **not only can the essay still be gamed, but it's easier than ever before.**

Whereas previously you might have struggled to come up with supporting examples for your stance, writer's block is hardly an issue on the new essay. Everything you need is in the passage. And since you're now given 50 minutes instead of 25 minutes, you have plenty of time to write a long, detailed response.

As one of the first guinea pigs to take the new SAT, however, I made some subtle mistakes that kept me from a perfect 8/8/8 essay. My initial framework was good but missed a few things. Since then, I've been able to refine and evolve my approach to incorporate what I've learned from my results as well as those of my students.

In this book, I'll lay out the approach that I've developed and show you what mistakes to avoid. To prove my advice is the real deal, I'll also share actual student essays that earned perfect scores as a result of the techniques I teach.

I hope you find these insights helpful.

How the SAT Essay is Graded

Before we get to how to write the essay, let's get some background information out of the way.

- The SAT Essay is optional, but most of the top 40 schools require it.
- It's 50 minutes long.
- It's given at the end of the test.
- The passage will be 650-750 words long.
- Though the passage is different from test to test, the assignment is always the same. It should never surprise you.
- You should use reasoning and textual evidence from the passages to support your analysis.

Once your essay is sent in, two readers each give it a score from 1-4 in three categories: Reading, Analysis, and Writing. The scores for each category are summed up.

	1st Reader 1-4		2nd Reader 1-4		Total 2-8
Reading	_____	+	_____	=	_____
Analysis	_____	+	_____	=	_____
Writing	_____	+	_____	=	_____

The three totals are then added up to get the final score, a value between 6 (the worst) and 24 (the best). This is the score colleges will look at.

Whereas previously your essay score was combined with your writing score, now the essay score is reported separately from the other sections and does not impact any of your other scores.

Students who are applying to top 20 schools should aim for at least a 21 out of 24 (at least a 7/8 in every category). If you follow everything I outline in this book, a 21 shouldn't be hard to obtain.

The grading rubric for the SAT Essay is available on The College Board's website, but because it's pretty vague and unhelpful, I'll reiterate only the points you need to know here. Future chapters will expand on these guidelines.

Reading—Did you understand the passage?

- Make sure to mention the main idea of the passage in your essay.

- Don't twist the author's argument.

Analysis—How well did you explain how the author builds his or her argument?

- Focus only on the parts of the passage that most contribute to the author's persuasiveness.

- Your essay should evaluate evidence, reasoning, and/or stylistic and persuasive elements in the passage. Note that you do not have to address all three. You may choose to discuss only the stylistic elements, for example.

- Do NOT take a stance on the issue. Your job is to analyze the author's argument, NOT show why you agree or disagree with the author.

Writing—How well did you write your response

- Break up your essay into discrete paragraphs (intro, body paragraphs, conclusion).

- Vary your sentence structures (short and long, simple and complex).

- Use some college-level words.

- Use correct punctuation.

- Maintain a formal style and objective tone.

Don't worry about keeping all these guidelines in your head. Yes, it's important to understand what the graders will be looking for, but we'll be giving you a template that takes care of everything for you.

Before we get to that template, however, we need to walk through the 7 elements of a perfect essay in the next chapter. There, we'll expand on some of these guidelines to make sure you know how to implement them.

3

7 Things Every Perfect Essay Must Have

Imagine that you're The College Board and after each and every test date, your mailbox gets flooded with hundreds of thousands of essays from students all across the world. How on earth would you manage to grade every single one of them in 2-3 weeks?

Well, you would hire a bunch of teachers who presumably know something about grading papers. You would also standardize the grading process so that scores remain consistent across the board. How? By training the graders to put their own opinions aside and base their scores solely on models that have already been set.

And to get through the sheer number of essays that must be read, you'd require them to be quick.

Well, that's pretty much The College Board's grading process.

The graders don't spend any longer than TWO MINUTES reading your paper. Two minutes and that's it. It's a snap judgment: *What does this essay look like? A 4/3/3. Next!* Any flashes of your literary brilliance will be glossed over in a ruthless grading procedure that only cares about whether your work matches the standard model.

Your job, then, is not to write a masterpiece. This isn't your AP English class. The teacher won't be looking at every word and dissecting all your punctuation marks.

Instead, your job is simply to write something that looks like a 4/4/4 (the highest score you can get from each grader). Write something that looks smart, something that looks like all the essays that have been awarded perfect scores before, and you in turn will be rewarded with a perfect score. That's it. It doesn't have to be innovative and it doesn't have to be in your voice and style. Just give them what they want.

Whenever I explain this mindset to students, there's always a slight outcry because it conflicts with the creative writing process they've been taught throughout school. A lot of students want to stand behind their own writing abilities instead of relying on any kind of pre-scripted formula. "I'm already a good writer," they say. "I don't need to follow a template."

I totally understand. As much as I would like the essay portion to be less of a game and more of an indicator of true writing ability, it's not. A 50-minute essay is hardly a good way to measure one's competence and the truth is, those who go in prepared with a good idea of what they're going to write do better than those who don't.

The SAT is simply not the place for you to get fancy. There will be plenty of opportunities later on, whether you want them or not, to express yourself more creatively. The SAT is not one of them. Your purpose is to ace it and get into college, not start the next great American novel.

Having said all that, let's take a look at the characteristics of a perfect essay.

1. Length

Is it possible for a short essay to receive a perfect score? Absolutely. Does that mean you should focus entirely on quality instead of quantity? Absolutely not.

Most students don't realize just how much essay length matters. While quality does count, you're much better off sacrificing some of that quality for a longer essay. In a 2005 *New York Times* article, writer Michael Winerip reports how Dr. Les Perelman, one of the directors of writing at MIT, posted student essays on a far wall and tried to predict their scores based solely on length. After he finished, he was stunned by the results—his predictions had been right over 90 percent of the time. The shorter essays received the lowest scores and the longer essays received the highest. More often that not, an essay's score was determined by its length.

With that in mind, you want to fill up as much space as you can, at least 2.5 pages out of the 4 pages they give you. Filling up 2.5 pages means you have to **write fast**. Read the passage and go. **You want to spend as many minutes writing as possible.**

Now this doesn't mean you should start lengthening the spacing between words. The graders are wary of essays with inflated spacing and I can tell you from my students' experiences that it definitely doesn't work. Write how you would normally but do so at a much faster pace. My essay in Appendix A shows the degree of illegibility you can get away with.

2. Thesis

As you probably have learned countless times from your English classes, a thesis is a statement that captures the main idea or essence of your essay. It's usually placed at the end of the introductory paragraph.

Always have a thesis statement because it's something graders can easily spot and reward you for. They should know where your analysis is headed after reading it. I'll be teaching you how to write easy and effective thesis statements in the next chapter.

3. Organization and Structure

Always make sure you use paragraphs and that each paragraph serves a purpose that is distinct from the others. In other words, each of your paragraphs should contain a different idea but one that still relates to the overall analysis.

Essays that are just one humongous paragraph don't get high scores, so make sure you indent!

4. Vocabulary

People judge others by the level of their vocabulary. If you've ever thought of a person more highly because of his or her ability to use words you didn't understand, then you probably know just how much vocabulary can impress people. Well, the SAT essay graders are no exception. After all, good essays are good in part because they have sophisticated yet well-chosen words. Later on, we will be showing you not only what words to use but also some automatic ways of injecting them into your essay.

On the flip side, avoid using vague words like "stuff", "things", "lots of", "many cases", "many reasons", especially when you don't clarify them later with specifics.

5. Sentence Variety

Good writers use a mix of simple and complex sentences. Essays consisting solely of simple sentences don't get high scores. Imagine reading a paragraph like this:

> People are most productive under pressure. They have to think faster. They work harder to meet deadlines. Most people are lazy. They need motivation. Pressure is the best motivator.

Choppy and repetitive—not what you want. Let's take a look at a revised version:

> People are most productive under pressure because they have to think faster and work harder to meet deadlines. Without motivation, most people are lazy. For those reasons, pressure is the best motivator.

Still not perfect, but a whole lot better. Notice that the main ideas didn't change, only the way in which they were connected.

By definition, the secret to creating complex sentences is **the comma**. By finding appropriate ways of inserting commas, you'll naturally start to add more detail to your sentences. Let's use an example sentence to see how this works:

> The tiger ate my aunt earlier today.

Now let's add in some phrases:

After starving for several hours, the tiger, which had previously been so well-behaved, ate my aunt earlier today.

Notice that by adding in those phrases, which, by the way, required commas, **we not only made the sentence complex but also made it more detailed, which is part of what good writing is all about.**

Here is a toolbox of things you can do for sentence variety:

Tool	Example
Use *because*	The tiger ate my aunt earlier today because it was hungry.
Use *who, which,* or *that*	The tiger, which had been very hungry, ate my aunt earlier today.
Use a conjunction (FANBOYS)	The tiger was hungry and ate my aunt earlier today.
Put *although/even though* in front	Although it is usually well-behaved, the tiger ate my aunt earlier today.
Use a modifier	Always on the lookout for food, the tiger ate my aunt earlier today.

As you practice, make it a point to put complex sentences among simple ones.

6. Transitions

Another weapon in the sentence variety toolbox is transitions, words that show how your ideas are connected. These are so important to getting a high score that they deserve their own section. If you look at the official essay rubric on the The College Board's website, it's clear that graders are required to look for them. Not only do they serve as the "glue" between your ideas, but they also give your sentences rhythm and structure.

Common Transition Words

Example	Transition...	Similar Transitions
I love eating vanilla ice cream. **However,** *too much of it makes me sick.*	presents an opposing point or balances a previous statement	*fortunately, on the other hand, conversely, whereas, while, in contrast*
Math trains you to approach problems more analytically. **Furthermore,** *it helps you calculate the exact amount of tip to be left for the waiter.*	adds new and supporting information	*in addition, also, moreover, and, too, as well, additionally, not to mention*
Pandas are rapidly becoming extinct. **In fact,** *some experts predict that pandas will die out in 50 years.*	gives emphasis to a point by adding a specific detail/case	*as a matter of fact, indeed, to illustrate, for instance, for example*
The state is facing a flu epidemic. **Consequently,** *all hospital rooms are filled at the moment.*	shows cause & effect	*as a result, because, hence, therefore, thus, as a consequence, accordingly*
Granted, *the SAT is a long and tedious exam, but it's necessary for college admissions.*	concedes a point to make way for your own point	*nevertheless, although, even though, despite, even if*
Place the bread on an ungreased baking sheet. **Finally,** *bake in a preheated oven for 10 minutes.*	shows order or sequence	*subsequently, previously, afterwards, next, then, eventually, before*
Social security numbers uniquely identify citizens. **In the same way,** *IP addresses identify computers.*	shows similarity	*similarly, likewise, by the same token*
In conclusion, *the world would be a happier place without nuclear weapons.*	gives a summary	*in all, to summarize, in sum, to sum up, in short, as mentioned, thus*

You'll want to include several transition words in your essay to show the grader that you understand their importance and how they're used. The essay template in this book will give you easy ways to add both transitions and sentence variety to your essay.

7. Quotes from the Passage

If you read any of the top-scoring sample essays on The College Board website, you'll notice that they all quote extensively from the passage.

Why is this the case?

Because to write an analytical essay, you must point out what you're analyzing before you analyze it. If, for instance, you're discussing a writer's great choice of words, it only makes sense to provide specific examples of those words before you talk about how they contribute to the article's persuasive power. How do you do that? By quoting from the passage.

Quotes are the foundation of every analytical essay. The essays that speak of the passage in general terms without citing any specifics are the ones that do the worst.

Of course, you should never include quotes just to beef up your essay. You must quote with purpose and with proper punctuation. Don't worry. We'll be covering everything you need to know in a future chapter devoted entirely to the art of quoting.

The Elements of Persuasion

In this chapter, you'll learn about all the elements of persuasion an author might use to build his or her argument. No passage will contain every single one, so it will be your job to identify which elements are present in the one you're given.

For each element, I've included an example of its usage as well as a summary of the effects it might have on the reader. You'll find these summaries extremely helpful once you start writing essays using the template in the next chapter.

I cover the elements that show up the most first. Word choice and statistics are used in nearly all the articles you'll read.

In addition, you'll often see multiple elements being used at the same time. For example, a rhetorical question might also contain an appeal to emotion. Even though I discuss each element separately, this overlap is completely normal and should give you even more to talk about in your analysis.

Finally, the example excerpts you see in this chapter have all been adapted from articles with the following main arguments:

- There has to be alien life on other planets.
- Animals should be treated as if they were people.
- Save the environment, not to save the Earth but to save ourselves.
- The United States should have a direct democracy, instead of a system of representatives.
- Girls make equally capable engineers as boys.
- Having a best friend is necessary for child success.

Rhetorical Elements

1. Word Choice

Word choice refers to using intense, lively, or thematically-similar words in a manner that leaves a certain impression.

Uses include:

- evoking emotions or images (imagery)
- characterizing a subject or topic in a particular way
- associating positive or negative connotations with something
- setting the tone

Example 1

Argument:	Animals should be treated as if they were people.
Excerpt:	There aren't enough differences between humans and animals to condone the widespread practice of **factory** farming, which differs from **concentration camp** conditions only in that animals are overfed rather than underfed, and **injected** with growth hormones rather than **gassed**, all so we can **butcher** them for more meat.
Analysis:	The author uses quite a few strong words to portray the deplorable way animals are treated. The words in bold evoke images of The Holocaust, effectively associating all the horror of Nazi Germany to factory farms.

Example 2

Argument:	Animals should be treated as if they were people.
Excerpt:	Dolphins **wave** to their trainers and **listen** for instructions. They'll then **hold a brief conference** underwater to **plan** their synchronized **jumps** through the hoops.
Analysis:	The author's main argument is that we should treat animals as if they were people. By using words that typically pertain only to people, the author is able to "humanize" dolphins in the reader's mind and hopefully evoke sympathy for them. This literary device is called anthropomorphism.

Example 3

Argument:	The United States should have a direct democracy, instead of a system of representatives.
Excerpt:	First someone submits a bill. Once it gets the requisite sponsorship in Congress, it goes to a committee. This **gaggle** of **supposedly enlightened** politicians then **dilly-dallies** over the fine print, putting whatever **mumbo jumbo** they want in the bill before it's voted on.
Analysis:	The author uses the words in bold to characterize politicians as incompetent, bumbling fools. This is especially persuasive because it implies that even the reader would be more effective in government than they are. By making readers feel good about themselves, the author is able to win them over.

2. Statistics/Data

Possible uses:

- to indicate a problem
- to make an idea hard to argue against because numbers are perceived as facts, not opinions
- to ground the author's argument in reality
- to surprise readers
- to put one quantity in relation to another for compare/contrast

Example 4

Argument:	There has to be alien life on other planets.
Excerpt:	The Universe is simply too vast, too filled with planets and stars, for us to be alone. The Milky Way contains 100 billion stars, including our own Sun. In the course of surveying thousands of stars over the last few years, the Kepler telescope has found that nearly all of them have planets and at least 17% have Earth-sized worlds orbiting them.
Analysis:	By noting the incredible number of stars and orbiting planets, the author makes a strong case based on sheer probability that there is life on other planets.

Example 5

Argument:	The United States should have a direct democracy, instead of a system of representatives.
Excerpt:	For example, each state gets two Senators so that all states have equal representation. This seems good until you consider that California has 80 times the number of people as Wyoming. **That means a single resident of Wyoming has the same Senatorial power as 80 residents of California combined.**
Analysis:	The bolded sentence shocks the reader and incites a sense of unfairness. By comparing Wyoming residents with California residents using an exact number, the author shows the extent to which our system of democracy is broken.

Example 6

Argument:	Girls make equally capable engineers as boys.
Excerpt:	Why does America have so few female engineers? More than 50% of American college students are now female, but only about 1 in 5 engineering faculty or tech startup founders are women.
Analysis:	The author uses statistics to point out not only a problem but also the extent of it. Readers are filled with a sense of inequity.

Example 7

Argument:	Girls make equally capable engineers as boys.
Excerpt:	Iran is not a women's rights haven. In this country, women can actually be stopped in public by actual morality police if their clothes are deemed too tight or revealing, and married women have to ask their husbands' permission to leave the country. Yet somehow, 70% of Iran's science and engineering students are women and so are a huge proportion of its tech startup founders. How did that happen?
Analysis:	Using the 70% figure, the author draws an implicit comparison between the United States and Iran, which has a much higher proportion of women in tech and engineering fields. If a country that limits women's rights can produce such great numbers of female engineers, then surely there is no excuse for the disparity between men and women in the U.S. This comparison effectively shifts the burden of proof to the skeptics.

3. Appeal to Authority

An appeal to authority usually consists of quotes from authority figures, research from respected universities, or actions of government or other authoritative bodies. It can

- raise credibility by showing the author is not the only one who believes in an idea
- increase trust by showing that an argument is well-researched
- gain acceptance or sympathy from readers who identify or think highly of the authority figure
- establish a precedent that puts peer pressure on other people to do the same

Example 8

Argument:	Animals should be treated as if they were people.
Excerpt:	In 2013, for example, the government of India declared dolphins to be "non-human persons." This made keeping them captive for entertainment purposes illegal everywhere in the country.
	And in 2015, a New York judge ruled that two chimpanzees could not be "unlawfully detained" for research purposes, citing the writ of *habeus corpus*. *Habeus corpus*, literally meaning "show me the body," states that no person may be detained without evidence of wrongdoing. In this ruling, the judge extended Constitutional protection for "persons" to chimpanzees.
Analysis:	The fact that places as far away from each other as India and New York City have passed rulings in support of the humane treatment of certain animals gives the author's argument credibility and establishes a precedent. In other words, the idea has already been accepted, and the pressure is on other governing agencies to do the same. It's easier to persuade people to do something when they won't be the first to do it.

Example 9

Argument:	There has to be alien life on other planets.
Excerpt:	World-renowned physicist Stephen Hawking agrees. "To my mathematical brain, the numbers alone make thinking about aliens perfectly rational. The real challenge is working out what aliens might actually be like."
Analysis:	By quoting Stephen Hawking, who is known and admired both inside and outside the physics world as a brilliant scientist, the author makes his argument much harder to dismiss. For readers to disagree with the author, they would also have to disagree with one of the world's most respected physicists.

4. Acknowledging the other side/Making Concessions

A concession is a point that is strategically given up or granted to the opposing side. Acknowledging the other side allows an author to

- address counterarguments, doubts, or fears readers may have

- establish common ground

- pave the way for new points to be made, instead of getting bogged down in old ones

Example 10

Argument:	The United States should have a direct democracy, instead of a system of representatives.
Excerpt:	Of course, we'd be dismantling the very system that America's founding fathers put in place. After all, the system of government we have now is based on the Constitution that they drafted. However, there were some logistical barriers to a direct democracy when the Constitution was written. There was, for example, no Internet. No way for every American to weigh in on a given issue in a timely fashion. But now there is. And if the founding fathers were alive today, they'd see the possibilities and advocate for a system that's more inclusive of everyone.
Analysis:	The author willingly acknowledges that a direct democracy would run counter to the Constitution. That way, he is able to address it and carve a way past a major obstacle to his argument. In doing so, he is able to frame America's founding fathers not as enemies but as allies. If the author did not acknowledge the other side, he would not be able to quell the doubts they may have.

Example 11

Argument:	Save the environment, not to save the Earth but to save ourselves.
Excerpt:	During the Permian Extinction, Earth got so hot that the land mass now known as Russia became one giant lava flow. I'm sure some skeptics of climate change will say, "See? The Earth survived that. Surely it can survive our puny little greenhouse gas emissions without a problem." Well, yeah. The Earth can survive alright. But what we selfish humans might want to know about the Permian extinction is that over 90% of all species living on Earth died out. That's why we don't see those cute little ammonites that you can only buy in fossil stores now, or those awesome fern-jungles you see in paintings of ancient Earth. They died out. We would have too, as I'm pretty sure skyrocketing global temperatures and global desertification would have killed all of our food crops.
Analysis:	The author establishes common ground with those who may not believe in saving the Earth by first agreeing with them: The Earth will survive whatever we do to it. However, this common ground only paves the way for his primary argument: Earth will survive, but we won't. By first identifying with his opponents' way of thinking, the author is able to weaken any resistance he may face in leading them down his own line of reasoning.

Example 12

Argument:	Having a best friend is necessary for child success.
Excerpt:	Many adults are now saying that children should not have best friends. In recent years, child psychologists have raised concerns about the exclusivity and potential for possessiveness in these relationships, which may detract from the varied social interactions important to childhood learning, or result in hurt feelings if one's best friend moves on. As a parent, none of this worries me at all. **Don't get me wrong—I've had my share of trials and tribulations with best friends whom I've had to say goodbye to. But these relationships, in good times and in bad, have given me invaluable practice for handling adult partnerships.**
Analysis:	The author identifies with the audience and concedes that she herself has had bad experiences with best friends. This acknowledgment allows her to make a very strong point—that even with those bad experiences, the benefits of having best friends outweigh the costs.

5. Analogies/Comparisons

Analogies are comparisons between two things. In general, they

- allow readers to understand more complex concepts by comparing them with simpler ones

- associate new ideas with ones the reader is already familiar with

- lead the reader into agreement by connecting something new with something the reader has agreed with or done

Example 13

Argument:	There has to be alien life on other planets.
Excerpt:	In 2010, analysis of data from the Mars Global Surveyor found that methane concentrations in the red planet's atmosphere increase during the warm season. Why is this intriguing? Because methane is created by bacterial life here on Earth. **And anyone who's ever taken out stinky garbage during the summer will tell you that bacterial life grows much faster—and produces much more stinky gas like methane—when it's warm.**
Analysis:	By comparing methane on mars to taking out the garbage, the author makes his point relatable and easier to understand. Readers must first understand something before they can be persuaded by it. Furthermore, this analogy strengthens the link between bacterial life on Earth and the possibility of bacterial life on Mars.

Example 14

Argument:	There has to be alien life on other planets.
Excerpt:	While naysayers point out that we've never received radio signals from another civilization despite scanning the skies for such signals for decades, Hawking has another explanation for the silence. He believes that a smart extraterrestrial civilization would hide its existence to avoid being attacked by others. **"If aliens visit us," Hawking says, "the outcome would be much as when Columbus landed in America, which didn't turn out well for the Native Americans." If we're not hearing from alien civilizations, it's probably because they've learned to keep their communications private, like any smart family in a crowded neighborhood.**
Analysis:	The analogies to Columbus and "any smart family in a crowded neighborhood" add validity to the author's main point and relates the idea to concepts readers are already familiar with. The comparisons make it seem like it's common sense that aliens would not want to contact us.

Example 15

Argument:	The United States should have a direct democracy, instead of a system of representatives.
Excerpt:	Some suggest that the risks of hacking and voter fraud are too great. Really? You do all your banking online and you're telling us the Internet is not secure enough for voting?
Analysis:	By comparing online banking to online voting, the author is able to downplay the risks of a direct democracy in which we all vote online. After all, if we already trust online systems to handle our money, there's no reason we shouldn't also allow them to handle our votes.

6. Juxtaposition

Juxtaposition is placing two things side by side for comparison or contrast. Unlike an analogy, it does not try to relate one concept to another. Instead, it merely positions them together in such a way that a significant distinction is highlighted or one option is made to seem better than the other.

Example 16

Argument:	Save the environment, not to save the Earth but to save ourselves.
Excerpt:	The idea of being "kind to the planet" assumes a couple of things. First, it assumes that we are capable of helping or harming Earth in some meaningful way. We're not. Earth has weathered asteroid impacts and climate change that would have blown us off the map a dozen times over, and Earth's ecosystems have always adapted. Human activity is nothing. Earth will bounce right back from any havoc we may cause. It always does. Humans, however, will not.
Analysis:	By juxtaposing Earth and humans, the author contrasts the fragility of humans with the durability of planet Earth. The intent is to make us feel small and insignificant, especially on a scale that includes "asteroid impacts" and "climate change." By making us feel this way, the author compels us to identify with her argument that we are most at risk when we don't save the environment, not Earth itself.

In SAT essay passages, you will often see authors juxtapose the past and the present, or one country with another, to emphasize a problem (e.g. In 1960, 83% of our clothing came from factories in the United States. By 2013, that percentage had declined to 18%.)

Example 17

Argument:	The United States should have a direct democracy, instead of a system of representatives.
Excerpt:	Who wants rich, privileged bureaucrats bossing us around when we can have people like you and me fighting for our own causes?
Analysis:	By juxtaposing "bureaucrats" with "people like you and me," the author creates an "us versus them" mentality. The implication is that politicians are totally out of touch with the common people they're supposed to serve. Furthermore, the reader is put into a situation with seemingly only two options, and of course, one is made out to be more enticing than the other.

7. Challenging Assumptions

By challenging common assumptions, an author

- enables an argument to proceed from a clean slate

- dismisses any preconceived ideas or biases that may run counter to his or her argument

Example 18

Argument:	Animals should be treated as if they were people.
Excerpt:	I've had to think twice about ordering a burger ever since I found out that cows have best friends. That's right—scientist Krista McLennan discovered in 2011 that cows pair off, and not in a mating sense. Instead, a female cow will develop an especially close bond with another fellow cow and show signs of stress if they are separated.
Analysis:	The author challenges the assumption that animals don't behave or feel as humans do. By moving the reader away from what is likely a widely-held belief, the author is able to argue from a clean slate and dismiss any preconceived notions that may sabotage the argument.

Example 19

Argument:	There has to be alien life on other planets.
Excerpt:	Recent experiments have shown that organisms can thrive with just two requirements: a temperature that allows for chemical bonding and an energy source such as the sun. Water is not necessary. In fact, certain proteins that usually contain water such as myoglobin can function just fine without it. And that's just Earth-based organisms we're talking about. There's no reason the biochemistry of an alien life-form can't run on something other than water.
Analysis:	The author tears apart the common knowledge that water is necessary for life. By doing so, he opens up many more possibilities for alien life to exist. The ability to persuade the reader is no longer constrained by a limiting assumption.

8. Anecdotes

Anecdotes are short stories, often personal, that are used to make a point.

Example 20

Argument:	Animals should be treated as if they were people.
Excerpt:	A 2008 study found that crows are able to distinguish one human from another and react differently depending on how they are treated. In 2011, a four year old Seattle girl named Gabi Mann dropped a chicken nugget only to have a crow swoop in to eat it. She soon realized that the crows were watching her, looking for another bite. As time went on, she began feeding them on a regular basis. That's when the gifts started appearing: a miniature silver ball, a blue paper clip, a black button, a yellow bead, and the list of shiny objects goes on. It's a peculiar collection of objects for a little girl to treasure, but to Gabi these things are more valuable than gold.
Analysis:	The author uses the anecdote of Gabi's crows to form a strong emotional bond between humans and animals in the reader's mind. By giving an illustration of how animals can act like humans, the author opens us up to accepting his main argument.

Example 21

Argument:	Having a best friend is necessary for child success.
Excerpt:	Growing up, I was always a bit of an odd duck. I was interested in things that other children weren't and bored by things they loved. I didn't really have a best friend I could relate to until the fifth grade, and when I did, it was so, so validating. We would squirrel ourselves away from rambunctious sports games to go for walks in the forest. We would get to school early to chat about our little hobbies. My bestie dragged me to places I wouldn't have ventured to alone. These experiences helped me grow into the person I am today.
Analysis:	The author shares some of her personal memories and experiences in order to convey the importance of having a best friend. Her story is one that nearly all readers will be able to relate to and find similarities with. By establishing common ground through this anecdote, she opens up a strong connection with the reader.

9. Rhetorical Questions

A rhetorical question is one that isn't answered by the author. Either the question doesn't need to be answered because the point being made is self-evident or it's designed for readers to answer themselves.

Possible uses:

- prods readers into agreeing or answering for themselves in a certain way

- gets the reader to imagine a certain scenario

- lays out common ground or assumptions that the author can then build upon

Example 22

Argument:	Save the environment, not to save the Earth but to save ourselves.
Excerpt:	To understand just how much temperatures rose during the Permian Extinction, consider this: have you ever been outside on a really hot day? Like, really hot. We're not talking about some wimpy 84° days. I mean like, 104° heat. Massive-epidemic-of-heatstroke heat. That's how hot the ocean was.
Analysis:	The author uses a rhetorical question to make the experience of a hot summer day more real and visceral. By reaching out to the reader's senses, the author makes a deeper impression when she later states, "That's how hot the ocean was."

Example 23

Argument:	The United States should have a direct democracy, instead of a system of representatives.
Excerpt:	**Democracy is the greatest system of government ever, right?** It's the only system where We The People get to decide what goes on in our land—not some overlords who rule us by force.
Analysis:	By ending the bolded sentence with "right?" the author is nudging the reader to respond mentally in the affirmative. A small gesture of agreement establishes common ground and give the argument momentum.

Example 24

Argument:	The United States should have a direct democracy, instead of a system of representatives.
Excerpt:	The maze of American politics is such that few Americans even understand who makes our laws—much less where those people come from. Doesn't that strike you as a problem?
Analysis:	The sharp rhetorical question compels the reader to answer in a frame of logic that has been set by the author. Obviously, the answer is "yes," but by leaving it up to the reader to answer in the affirmative, the author is able to induce readers to persuade themselves. After all, there's nothing more persuasive than the thoughts and beliefs we ourselves come up with.

Example 25

Argument:	The United States should have a direct democracy, instead of a system of representatives.
Excerpt:	So laws in America today are made by elite politicians who are elected through a convoluted system that doesn't represent anyone. What if We The People just made the laws?
Analysis:	The rhetorical question at the end prompts readers to imagine an ideal scenario in which they are the ones who make the rules. Even if that scenario is far removed from reality, the author is able to persuade just by putting the possibility into the minds of readers, and they are compelled to at least think about the issue for themselves.

In general, stay away from analyzing rhetorical questions as a stand-alone persuasive technique on the SAT Essay. Why? Because there's typically only one or two rhetorical questions in any given article, and that's just not significant enough for an entire paragraph's worth of analysis. Instead, focus on rhetorical elements that occur throughout the article, not just in one sentence.

However, if the author makes heavy use of rhetorical questions or they work alongside another persuasive element, then mentioning them is a good move. For example, you will often see juxtaposition or a challenging of assumptions worked into rhetorical questions themselves.

10. Hypothetical Situations

The hypothetical situation is almost like a sibling to the rhetorical question because the uses are much the same even though it's not in question form.

Possible uses:

- Gets the reader to imagine certain possibilities without having to state them explicitly
- Allows the author to describe certain outcomes in a way that benefits his or her argument

Example 26

Argument:	Save the environment, not to save the Earth but to save ourselves.
Excerpt:	The Earth became scorchingly hot at least once before, and there's no reason it can't happen again. If it does, we could go the way of the dinosaurs and become popular little fossils for the hyperintelligent life-forms of the future to buy in jewelry stores.
Analysis:	The author puts forth the hypothetical scenario of the Earth becoming too hot for humans to bear. In doing so, he gives himself an opportunity to describe the catastrophe that may occur if we don't protect the Earth. If he didn't bring up this hypothetical situation, he wouldn't have this additional avenue of persuasion.

Example 27

Argument:	Having a best friend is necessary for child success.
Excerpt:	Imagine a world without the best friend you have now. Wipe out all those trips you went on together, the birthday parties you celebrated, and the long chats you had over the phone.
Analysis:	The author paints a stark hypothetical situation in order to win readers over emotionally. They are forced to think about the author's argument on a personal level, in terms that matter to them.

11. Appeal to Identity

An appeal to identity is one that takes advantage of the common values and beliefs of a group. It's persuasive because human beings are social creatures that seek belonging, and we gravitate towards those ideas that enhance that sense of belonging.

Example 28

Argument:	The United States should have a direct democracy, instead of a system of representatives.
Excerpt:	What is *really* stopping us from bringing direct democracy to America? I would argue that it's simply fear of change.
	Fear of change didn't stop our forefathers from crossing an ocean and settling a new continent. It didn't stop the Freedom Riders from risking their own lives in pursuit of equality for all. It isn't what invented the Internet or put smartphones in our pockets.
	As Americans, we as a people have always embraced change. And we deserve a change that will put the power of a truly democratic society in our own hands.
	If we want to see this dream become a reality, we must act. After all, the wealthy politicians of Washington are not going to be the ones to put themselves out of power.
	So let's start our petitions. Let's put it on our ballots.
	Let's embrace direct democracy, together.
Analysis:	Throughout this entire excerpt, the author makes a strong appeal to the American identity. She mentions our forefathers and the Freedom Riders to stir up our nationalistic pride. She also brings up key words and ideas that resonate with every true American—*democracy, change, pursuit of equality*. In doing so, she's able to frame the idea of a direct democracy as one that upholds American values.

12. Strong Directives with the Collective Pronoun "We"

Directives are just another name for a strong suggestion or command, such as "Let's grab pizza!". Not only do they inform the reader of the next steps to take but they are also a call to action.

Typically, directives are used with the collective pronoun "we". Why is "we" significant? Because it serves to connect the author and the reader as being part of a larger group with a common cause. By using "we," an author portrays him or herself as being on the same side as the audience, one who will stand beside them in unison.

We'll use the previous example to illustrate.

Example 29

Argument:	The United States should have a direct democracy, instead of a system of representatives.
Excerpt:	What is *really* stopping us from bringing direct democracy to America? I would argue that it's simply fear of change.
	Fear of change didn't stop our forefathers from crossing an ocean and settling a new continent. It didn't stop the Freedom Riders from risking their own lives in pursuit of equality for all. It isn't what invented the Internet or put smartphones in our pockets.
	As Americans, we as a people have always embraced change. And we deserve a change that will put the power of a truly democratic society in our own hands.
	If we want to see this dream become a reality, we must act. After all, the wealthy politicians of Washington are not going to be the ones to put themselves out of power.
	So let's start our petitions. Let's put it on our ballots.
	Let's embrace direct democracy, together.
Analysis:	The bolded sentences serve to unify the audience and establish a common base of American values. The sentences toward the end are calls to action that incite impassioned readers to work towards a direct democracy.

Evidence

Though this chapter focuses mainly on the rhetorical elements you should know, that's not all your essay is limited to. The author's use of evidence is also something you can discuss. I'll go over two common types of evidence here: results from research and supporting examples.

13. Results from Research/Studies

Example 30

Argument:	Animals should be treated as if they were people.
Excerpt:	In McLellan's case, her team measured the heart rates of cows, which, like in humans, go up when the animals are feeling stressed, and the levels of cortisol, a stress hormone, in their blood. Sure enough, when cows were separated from their best buddy, their heart rates went up and so did their cortisol levels, almost as if they were worried about their best friend's absence.
Analysis:	By drawing upon the results of an experiment, the author solidifies her argument on a scientific basis, which builds more credibility and reinforces the similarities between humans and animals.

14. Supporting Examples

Example 31

Argument:	Save the environment, not to save the Earth but to save ourselves.
Excerpt:	You might be thinking that I'm blowing this whole "climate change" thing out of proportion. After all, you've barely noticed any changes in the weather, right?
	Unless, of course, you live in one of the U.S. states that has experienced the polar vortex, where changes to convection currents due to a warming ocean led to weeks of $-40°$ temperatures.
	Or you're a resident of California, whose nearly empty aquifers have led to extreme water rationing measures in some of the biggest population centers in North America.
	Or you're really hoping that those tropical bugs that carry Zika and Chagas disease and malaria will stay around the equator where they belong, and not spread north as they have been for the past few decades.
Analysis:	The author provides more than one example of the detrimental effects of climate change to magnify the extent of the problem. The variety gets readers to think of climate change not as an isolated problem but as one that will eventually affect them if they don't do something about it. Indeed, the author clearly wanted to make the issue of climate change loom large in the reader's mind. The examples themselves describe dire circumstances that elicit fear and despair.

Logos, Pathos, and Ethos

Logos

Logos is an appeal to logic or reason. Using statistics, challenging assumptions, and bringing up research are typical ways of making a logos-based argument.

Pathos

Pathos is an appeal to emotion (pity, sadness, greed, fear, joy, ambition, etc.). While using personal anecdotes or emotionally-charged words are common ways of invoking pathos, what an author chooses to discuss (the actual subject matter) is the primary weapon of pathos. Charities hoping to raise money for developing countries describe infant children who die in the arms of their mothers because they know it inspires guilt, pity, and compassion. Companies selling home alarm systems overplay the need to protect your family to make you buy out of fear. A CEO who wants company employees to work harder might stoke their ambitions by telling them their work is changing the world.

Ethos

Ethos is an appeal to the author's own identity, character, or trustworthiness. Think of a presidential candidate describing his accomplishments and military service to appeal to voters.

I put logos, pathos, and ethos at the end of the list of rhetorical elements because I consider them over-arching terms of persuasion rather than ground-level techniques. Most of the time, what you see as logos, pathos, or ethos can actually be broken down into other elements of persuasion (e.g. statistics are typically a component of a logos-based argument). When that happens, I prefer to discuss the ground-level technique (i.e. statistics) as the persuasive element rather than the more abstract Greek term (i.e. logos).

In cases when I do want to bring up logos, pathos, or ethos in my analysis, I like to refer to them not by their names but by their more natural sounding definitions. Rather than saying that an author is invoking pathos, for example, I might instead say that he's appealing to emotion or fear. I've found that while using a Greek term in your essay might seem impressive, it can actually look forced and contrived if done in the wrong way. However, this is just my personal style. You are certainly welcome to mention logos, pathos, and ethos as modes of persuasion in your essay. In fact, I've seen many high-scoring essays whose paragraph structure is based entirely on these three concepts (i.e. logos for body paragraph 1, pathos for body paragraph 2, and ethos for body paragraph 3). You'll see some examples of how to write about logo, pathos, and ethos in the student essays I share later in this book.

After reading this chapter, you may have noticed that a lot of the examples showcase more than one technique. This is not uncommon. Rhetorical questions will often pose hypothetical situations. Statistics may be used in conjunction with word choice. An appeal to identity can also appeal to emotion.

This overlap is good because it gives you more to talk about.

The way to handle overlap is to focus on one main persuasive element for each paragraph and use the other elements as additional reasons for why something in the passage is persuasive. Don't worry if this is a bit vague or confusing. You'll get a better sense of how to put everything together in the next two chapters.

5

The Essay Template for any Prompt

Now that you've learned what constitutes a perfect essay and the persuasive elements you'll be looking for, this chapter will introduce a repeatable framework that will make essay writing easy. After all, an essay is more than just analysis; there needs to be an underlying structure that organizes it. This is THE template that my students have used to get perfect scores. It may seem complicated and hard to memorize at first, but if you actually practice it a few times, you'll see how easy it is to produce amazing essays.

As you go through the template, notice the transitions and complex sentences that have been integrated by design. They ensure that your essay is built on a solid foundation.

Introduction

The idea that   Main argument of the passage   has its roots in   Noun   but the underlying reasons are often overlooked.

In   Title of passage , author   Author of passage   puts forth a detailed argument for   Noun .

In doing so, he/she employs a variety of rhetorical elements to persuade the reader, including

  Persuasive element 1 ,   Persuasive element 2 , and   Persuasive element 3 .

Body Paragraph 1

Author of passage 's deft use of Persuasive element 1 begins with his/her discussion of Noun .

Your analysis: Narrate & Explain (covered in the next chapter)
6+ sentences

Although Persuasive element 1 may come off as hyperbole/pandering/overly dramatic,

Your rebuttal (reinforce why it's effective) .

Body Paragraph 2

Just as persuasive as his/her use of Persuasive element 1 is Author of passage 's Persuasive element 2 .

Your analysis: Narrate & Explain (covered in the next chapter)
6+ sentences

Without Persuasive element 2 , Author of passage 's message would lose Noun .

Body Paragraph 3

Lastly, Author of passage bolsters his argument by using Persuasive element 3 .

Your analysis: Narrate & Explain (covered in the next chapter)
6+ sentences

Conclusion

In summary, Author of passage —using Persuasive element 1 , Persuasive element 2 , and

Persuasive element 3 –effectively makes the case that Main argument of the passage .

It's his/her use of persuasive elements that not only inform the reader of the problem but also spur the reader into action.

At this point, you're probably very confused and overwhelmed. Don't worry! Using my very own essay, I'll illustrate exactly how this template is used. First, I'm going to be honest—I'm not a naturally talented writer. I was a solid B student in most of my high school English classes. But by using the template above, I was able get a 7/7/7 on the May 2016 essay with a few minutes to spare (I'll share the mistakes I made in a later chapter). The following is the typed-up, word-for-word version of my essay. See Appendix A for a copy of the original hand-written version.

Even though I can't reproduce the passage here (copyright issues), you should be able to spot each component of the template in this essay. Some components are a little out of order, but it's all there. If you'd like to read the passage, search the Internet for "Viewpoint: Air-Conditioning Will Be the End of Us" by Eric Klinenberg.

The idea that we should reduce our dependency on fossil fuels has its roots in environmentalism but its underlying reasons are often overlooked. In "Viewpoint: Air-Conditioning Will Be The End of Us," author Eric Klinenberg puts forth a detailed argument that society as a whole must reduce its consumption of air-conditioning to thwart climate change. In doing so, he employs a variety of literary elements and supporting examples, including statistics, concessions, and word choice.

Klinenberg's deft use of statistics begins with his discussion of air-conditioning use in America in paragraph one. He writes, "Today Americans use twice as much energy for air-conditioning as we did 20 years ago," further noting that this is "more than the rest of the world's nations combined." This statistic grounds his argument in reality, so that even skeptical readers will not be able to dismiss him so easily. The juxtaposition of America's energy consumption with not only its past but also the rest of the world indicates the intensity of the problem. The implication is that we have drastically increased air-conditioning useage beyond what is reasonable. By using these particular statistics, Klinenberg sounds the alarm on our energy use, signalling to readers just how rampant and excessive their habits of consumption may be. Later, in paragraph 5, Klinenberg brings up yet another statistic—sales of air conditioners rose 20% in India and China last year. Here, he makes it explicit that air-conditioning is not a problem distinct to America, but a global one.

Just as persuasive as Klinenberg's use of statistics is his relatability and acknowledgement of the other side. In paragraph 2 and 3, he identifies with those who may be hesitant to accept his argument: "I'm hardly against air-conditioning," he asserts. Indeed, he continues to enumerate may advantages of air-conditioning, which include saving "the lives of old, sick and frail people" and enhancing "productivity in offices." By granting these concessions, Klinenberg establishes credibility as someone who has thought through his point of view. Readers are more likely to perceive him not as an extreme environmentalist hell-bent on cutting off the luxuries of AC but as someone who realizes moderate air-conditioning has its place. Therefore, they are likely to be more receptive to his stance, knowing that their potential objections have been heard. Klinenberg smartly frames himself as an advocate of "public health programs," further creating trust with his audience by putting the greater good first.

Lastly, Klinenberg bolsters his argument with his specific word choice. In paragraph 3, he characterizes the damage caused by air-conditioners as "irreversible." In paragraph 4, he portrays stores in New York City as "pumping arctic air" while "burning through fossil fuels in suicidal fashion." By using words such as "arctic" and "suicidal," he invokes an intense kind of imagery, one that dramatizes the excessive way in which we consume AC. Indeed, we turn our "homes, offices, and massive commercial outlets into igloos." Klinenberg's message to the reader is loud and clear: our "burning through fossil fuels" is extreme, superfluous, and even ridiculous. There is simply no need to use up energy at the level we are doing so today. Although some may say that Klinenberg's choice of words borders on hyperbole, it is an effective way of turning readers' attention to an issue that likely applies to them on a daily basis. It reinforces the point that our own energy habits have left us in a dire situation that will only get worse if changes aren't made.

In summary, writer Eric Klinenberg—using statistics, acknowledgements to the opposing view, and specific word choice—makes the case that rectifying climage change requires a substantial reduction in air-conditioning usage. It's those persuasive elements that not only inform the reader of the problem but also spur the reader into solving it.

Step-by-step

Notice how closely my essay follows the template prescribed in this chapter. There is absolutely no way I would've been able to crank out that essay in 50 minutes if I didn't already have a solid blueprint to follow. Here's my step by step process after I first open the writing booklet:

1. Read the entire passage from beginning to end, underlining any words or sentences that contribute to its persuasive power. Next to each part you underline, quickly jot down the persuasive element being used.

2. When you finish the passage, go over the parts you underlined and make a mental list of the persuasive elements you identified.

3. From that list, choose three persuasive elements that you'll discuss in your essay. Try to choose ones that are used throughout the essay so that you'll have a lot to talk about. Don't choose persuasive elements that are used only in one sentence.

4. Read the essay task. It will tell you exactly what the author's main argument is.

5. At this point, it's time to begin writing your essay. You should be onto your introduction within 10 minutes of the starting time. That gives you at least 40 minutes to complete your essay.

6. Always always have a conclusion. The graders will punish you if you don't have one. I'd rather cut one of my body paragraphs short than run out of time before the conclusion. It's completely scripted out for you in the template so just bang it out when you have 3-4 minutes left. Like the introduction, the conclusion should be muscle memory.

The beauty of having a template is that it takes the thinking away from all the essay components that relate to organization, sentence variety, vocabulary, and transitions, allowing you to focus your time and attention on your analysis.

This template makes a low score nearly impossible because even if your analysis isn't the strongest, you'll still score well on organization and language use.

Some Frequently Asked Questions

1. *How do I know whether the author is a he or a she? Sometimes the names are gender neutral.*

   You can tell from the assignment box at the end of the passage, based on whether it uses *his* or *her*.

2. *I've been taught all my life to write an awesome "hook" that draws the reader in. Doesn't the hook in the template seem a bit generic?*

   First of all, the template's introductory sentence is amazing because it reiterates the passage's main idea and leads into your essay in a clear and expressive way. No other type of hook will impress the graders more. Remember that they're hired to rush through hundreds of essays full of poor writing from students everywhere. There is NO hook in the ENTIRE world that would excite them. I'm a teacher myself and I've read hundreds of essays from students taking practice tests for the first time. The best they've been able to get from me is a chuckle (usually from a terrible hook). Maybe I'm just a mean, humorless teacher, but still …

   Don't go out of your way and waste 5 minutes coming up with an awesome hook. Even if you manage to come up with an ingenious attention-grabber, it's just one sentence in your entire essay. The hook alone will not get you a perfect score. Writing a long intelligent essay will.

3. *Will the graders know I'm using a template?*

   First, realize that the template does not script out the meat of your essay, which is your analysis. Much of what you write will actually be unique to you. Secondly, graders look at so many essays that it's extremely unlikely they'll be able to tell whether you're using a template. None of my students have ever been penalized for using the template in this book (or any other template). Even if the graders somehow knew you were using this specific template from this specific book, they would still have to base their grades on the essay content and NOT on whether they think you used a template. If you're still concerned, then I recommend developing your own introductory sentence. After all, the first sentence is more likely to tip a grader off than anything else in the template. This leads me to the next question …

4. *Can I tweak the template to fit my own style/needs?*

   Absolutely. Everything in the template is a guideline. You can use as little or as much of it as you want. The key is to have some default nuts and bolts already in place so you can focus on implementing the rest of the concepts in this book. The stronger you are as a writer, the less you need to rely on a template. Later on, I'll be showing you some real perfect-scoring essays that showcase what I teach. Some of them follow the template and some don't.

5. *Do I need to practice?*

   Um yes! Keep writing essays until you can consistently churn out high quality responses. Do not think that it'll come easy just because you read this chapter. So many students, including my younger self, think that they can pull off a great essay just by reading some formula or template. Then when test day inevitably rolls around, they freeze up and totally botch it. Having the template in your mind is not the same as being able to write it. You need to make it muscle memory. Go and practice.

Narrating & Explaining

The previous chapter laid out the best approach to the SAT Essay, but there are still some things that haven't been answered. Namely, how you should fill out each body paragraph.

The template requires that each body paragraph have at least 6 sentences of analysis. In this chapter, I'll show you how to write your analysis using "narrating" and "explaining" sentences.

Overview

Here is an example of narrating and explaining in action.

> [1] Harken's deft use of statistics begins with his discussion of our voting system. [2] He mentions that "a single resident of Wyoming has the same Senatorial power as 80 residents of California combined." [3] By comparing Wyoming residents with California residents using an exact number, the author shows the extent to which our system of democracy is broken. [4] Such a lopsided statistic unsettles the reader and incites a sense of unfairness. [5] After all, our system of government is supposed to promote American values of fairness and equality, not the extreme opposite. [6] This implied betrayal of trust carries a strong emotional appeal and is sure to provoke a reaction from an American audience.

The 1st sentence is just the topic sentence for the first paragraph in the template.

The 2nd sentence **narrates**, or recounts something the author does. Think of a broadcaster narrating the action of a basketball game. A narrate-sentence consists of either a quote from the passage to point something out or a description of what the author is doing at a certain place in the passage.

Sentences 3 and 4 **explain** why the narrated portion is persuasive. What's the purpose behind writing it? What does the author hope to achieve? What effect does it have on the reader?

The 5th sentence also **explains**, but does so by *reiterating*, or summarizing, the author's message in a way that not only shows how it's persuasive but also sets up the next explaining sentence. Reiterating the author's message is a technique we'll explore in greater detail later in this chapter.

The 6th sentence piggy-backs off of the 5th sentence and further **explains** the effect on the reader.

The point of this framework is to ensure you have a purpose for each sentence in your analysis. As you can see, you'll typically narrate first and then explain, and the majority of your analysis should be explanation sentences. However, don't worry about how many sentences of each type you need to include. It depends on your analysis. For example, you might narrate for two sentences, explain for the next three, narrate again for one sentence, and explain for the next two.

The idea is to keep narrating and explaining until you fill out your paragraphs. So let's say you wanted one of your body paragraphs to focus on the author's use of word choice. You would pick out an example from the passage, narrate it, and explain it. Then you would pick out another word choice example from the passage and repeat the process. After that, you might conclude your paragraph or do another round of narration-explanation if you thought it would bring something new to your analysis.

How to Narrate

Narrating involves pointing out something the author says or does in the passage. To narrate effectively, you must be able to quote effectively. However, quoting is big enough of a topic that we have to reserve it for the next chapter. What we will do is go through the most useful narration techniques and phrases.

1. Quote a complete sentence from the passage

Example 1

Excerpt:	NASA's technology base is largely depleted. Currently, available technology is insufficient to accomplish many intended space missions.
Analysis:	The author emphasizes the fact that "available technology is insufficient to accomplish many intended space missions."

For a detailed analysis, sometimes it's necessary to quote a complete sentence or a long portion of one. However, try not to do this too often because it can take up too much space and come off as lazy. It doesn't look good when half your paragraph is a quote from the passage. If you see that happening, use another technique instead.

2. Weave words and phrases from the passage into your own sentence

Example 2

Excerpt:	For centuries the world's waters have connected us. Explorers, traders, scientists and fishermen have traveled our oceans and rivers in search of new resources and a greater understanding of the world.
Analysis:	Laura Bush points out how bodies of water have "connected us" and given us "a greater understanding of the world."

3. Describe what the author is doing without using quotes

Example 3

Excerpt:	Unfortunately, paid leave is facing significant opposition from some business interests, just as FMLA did 21 years ago. Back then, opponents said FMLA would be bad for business, yet according to a 1998 survey, just 5 years after the law was implemented, 84% of FMLA compliant businesses reported either no costs or actual cost savings as a result of their family and medical leave policies.
Analysis:	Gillibrand rounds out her argument by drawing a comparison between the modern fight for paid leave and the adoption of the Family and Medical Leave Act (FMLA) decades ago.

Example 4

Excerpt:	To remain the leader in aerospace technology, we must continue to perform research and invest in the people who will create the breakthroughs of tomorrow, preserving a critical component of our nation's economic competitiveness for future generations.
Analysis:	Braun warns that our investment in space will be a determining factor in America's potential to innovate.

Example 5

Excerpt:	America is not alone in its efforts to preserve marine treasures. Australia has expanded its protection of parts of the Great Barrier Reef and the United Kingdom announced the designation of the Chagos Islands Marine Reserve in 2010.
Analysis:	Bush mentions several countries that have made ocean conservation a priority, such as Australia and the United Kingdom.

Here are some helpful phrases to use when narrating. There's no need to memorize them but hopefully they give you a good sense of how to word these types of sentences. By exposing yourself to a variety of words, you can avoid repeating the same ones in your narration sentences.

The author...

- discusses
- frequently uses
- brings up
- cites
- writes
- expresses
- establishes the fact that
- indicates that
- argues that
- makes the conclusion that
- cites

- represents this issue as
- mentions
- notes that
- calls upon
- voices his concern that
- tells the reader that
- shows that
- continues
- characterizes
- warns
- points out

- quotes
- acknowledges
- contends
- states
- maintains
- emphasizes
- highlights
- makes a reference to
- asserts that

How to Explain

Remember that "explaining" sentences shed light on why something is persuasive. They focus on the *why* instead of the *what* and should make up the bulk of your analysis. There are several techniques you can use to explain how an author builds his/her argument:

1. Present the author's reasons

Why did the author write it? What were his/her intentions? Put yourself in the author's shoes.

Example 6

Excerpt:	We get that money is tight, we understand that there is a hierarchy of needs, and that the French Market or a Mark Twain plaque are not hospital beds and classroom size.
Analysis:	Through her concessions, Smith gains trust and portrays herself as a knowledgeable, objective advocate that is trying to find the best solution for everyone, not just herself.

Example 7

Excerpt:	The best available research, however, shows that previous increases in the minimum wage haven't decreased or increased hours or jobs in any statistically significant manner.
Analysis:	By citing established research, the author adequately addresses the objection that raising the minimum wage would lead to greater unemployment.

2. Walk through the reader's emotions

What is the impact on the reader's emotions? How does the reader become convinced of the author's argument? Consider the target audience's needs, desires, and fears.

Example 8

Excerpt:	We are at risk of permanently losing vital marine resources and harming our quality of life. Overfishing and degrading our ocean waters damages the habitats needed to sustain diverse marine populations.
Analysis:	By using collective pronouns, the author makes readers feel like the potential losses are their own losses.

Example 9

Excerpt:	In fact, compared to paper bags, plastic grocery bags produce fewer greenhouse gas emissions, require 70 percent less energy to make, generate 80 percent less waste, and utilize less than 4 percent of the amount of water needed to manufacture them.
Analysis:	These advantages over paper bags are likely to surprise readers and make them question their own assumptions about plastic bags.

Example 10

Excerpt:	At best, the Arctic Refuge might provide 1 to 2 percent of the oil our country consumes each day.
Analysis:	Carter specifically casts these figures as small percentages to influence the way we perceive the issue. One to two percent just doesn't seem significant in readers' minds.

Example 11

Excerpt:	One of the most unforgettable and humbling experiences of our lives occurred on the coastal plain. We had hoped to see caribou during our trip, but to our amazement, we witnessed the migration of tens of thousands of caribou with their newborn calves ... Standing on the coastal plain, I was saddened to think of the tragedy that might occur if this great wilderness was consumed by a web of roads and pipelines, drilling rigs and industrial facilities.
Analysis:	Jimmy Carter's story is one that the audience can relate to. After all, who hasn't been awed by the grand beauty of nature? Whether it's looking up at the night sky or hiking up a mountain, these experiences are something that readers cherish and are afraid to lose. In this way, Carter taps into the reader's fear of loss and establishes a common ground for the audience to empathize with his views.

3. Describe the purpose in the context of the author's argument

In what way does it help the author's argument?

Example 12

Excerpt:	As President Obama said in his State of the Union address last month, a woman "deserves to have a baby without sacrificing her job. A mother deserves a day off to take care of a sick child or a sick parent without running into hardship."
Analysis:	The quote from President Obama further substantiates the author's description of struggling women in the workplace.

Example 13

Excerpt:	Currents in the Pacific have created a plastic garbage dump twice the size of Texas.
Analysis:	This figure expresses the magnitude of the pollution problem in the form of a familiar quantity—the large size of Texas. This makes Bush's argument less abstract and more concrete.

4. Reiterate the author's message in a way that helps you explain how it's persuasive

Boil down what the author is trying to convey. You can reveal an author's strategy by getting to the heart of his message.

Example 14

Excerpt:	If we don't protect our oceans, we could witness the destruction of some of the world's most beautiful and important natural resources. Fortunately, Yellowstone offers a blueprint for protecting our oceans. President Ulysses S. Grant created Yellowstone National Park in 1872 at a time when large wild areas on the frontier were at risk. The founding of Yellowstone sparked a 50-year period during which many of the national parks we enjoy today were created.
Analysis:	By using Yellowstone National Park as a reference point, Bush makes her own proposal seem less of a radical idea. **Because Yellowstone was so successful, similar modern-day efforts could also be successful.**

Notice how the bolded sentence gets to the core of the author's message and clarifies our analysis in the previous sentence.

Example 15

Excerpt:	Knowledge provided by weather and navigational spacecraft, efficiency improvements in both ground and air transportation, super computers, solar- and wind-generated energy, the cameras found in many of today's cellphones, improved biomedical applications including advanced medical imaging and more nutritious infant formula, and the protective gear that keeps our military, firefighters and police safe, have all benefitted from our nation's investments in aerospace technology
Analysis:	By giving real-world applications of space research, Braun illustrates the importance of continued investment in our space program. **Life simply would not be the same without NASA.**

Here, the bolded sentence sums up the author's point to add emphasis to our analysis.

Don't confuse this technique with "narrating." Whereas "narrating" is pointing something out, "reiterating" is directly rephrasing it in a way that clarifies your analysis. When used sparingly, it can be a nice way of setting up or extending one of your points, but don't overuse it. Otherwise, you'll fall into the trap of summarizing instead of analyzing. Remember that your main job isn't to summarize the passage but rather dissect the persuasive elements within it.

Not all of the techniques above will apply in every situation. You'll have to choose which ones to use based on what you're discussing.

Putting everything together

Even though we covered narrating and explaining as separate topics, we only did so because it made them easier to teach. In reality, you'll have a lot of sentences that both narrate and explain. For example,

> Braun uses the words "endangered," "stark," "depleted," and "insufficient" to instill a sense of despair and urgency over NASA's decline.

Here, we're pointing out specific words an author uses and explaining their purpose all in the same sentence. These "combo" sentences are only natural. Narrating and explaining are just concepts to help you generate a detailed analysis, not separate categories into which you must force all your sentences.

The following is a list of my favorite sentence patterns that combine narration and explanation.

Pattern 1

Pattern:	By ..., [the author] [verb] ...
Example:	By repeatedly using the word "we", Smith closes the distance between author and reader.
Example:	By juxtaposing "bureaucrats" with "people like you and me," Rubin creates an "us versus them" mentality.

Pattern 2

Pattern:	Use a colon to pair your analysis with a quote.
Example:	Jones uses a surprising statistic to grab our attention: "Ninety percent of all Americans live within 15 minutes of Walmart."

Pattern 3

Pattern:	Set a quote off with dashes.
Example:	Although Smith provides only one statistic—"British libraries received over 300 million visits last year"—it's an important one that demonstrates the demand for libraries.

When writing these types of sentences in your own essay, you might find it helpful to draw upon the reasons for using each rhetorical element in "The Elements of Persuasion" chapter. The examples in that chapter are good models to follow.

Finally, here's an example body paragraph annotated in terms of narration and explanation so you can see how everything fits together in an essay.

Gioia's deft use of statistics begins with his discussion of the arts. **(Narrating topic sentence)**

He brings up a study that measured the drastic fall in arts participation in America. **(Narrating without a quote)**

He specifically draws attention to the fact that the "declines have been most severe among younger adults." **(Narrating with a quote)**

By citing these trends, Gioia is grounding his argument in reality so that even skeptical readers won't be able to dismiss it so easily. **(Explaining/Author's reasons)**

Not only do the numbers indicate that there is in fact a growing problem but they also show the extent of it. **(Explaining/Describing the purpose)**

The state of reading in America is in dire straits and because young people participate the least, the troubles will only compound. **(Explaining/Reiterating)**

The underlying assumption that young people are the future of America makes the statistics even more alarming. **(Explaining/Effect on reader)**

Gioia further points out that in a poll conducted in 2001, "38 percent of employers complained that local schools inadequately taught reading comprehension." **(Narrating with a quote)**

This additional statistic addresses any doubts readers may have over the value of literature engagement. **(Explaining/Describing the purpose)**

Certainly, literature is important from a cultural standpoint, but Gioia proves that it's also important from a practical standpoint. **(Explaining/Reiterating)**

By covering both bases, he is able to persuade readers with differing perspectives and values. **(Explaining/Author's reasons)**

Although Gioia's use of statistics may come off as overly dramatic, it is hard to argue with numbers that paint a bleak picture of an issue that most Americans are familiar with, whether it be in school or in the workplace. **(Explaining concluding sentence)**

How to Quote

Here are the rules you need to know when quoting from a passage.

1. Use quotation marks whenever you're quoting word-for-word directly from the passage

Example 1

Excerpt:	The panda could go extinct within the next two decades.
Sentence:	To convey a sense of urgency, Kingsley writes, "The panda could go extinct within the next two decades."

Notice the comma after *writes*. When a quote is preceded by an identifier such as *she said* or *according to experts*, a comma should be used after the identifier.

2. You should not use quotation marks when paraphrasing

Example 2

Excerpt:	The panda could go extinct within the next two decades.
Sentence:	Kingsley believes that the panda species may not exist in twenty years.

Not using quotation marks doesn't mean you can't borrow some words. It comes down to judgment, but as long as most of the words are your own, you don't have to use quotation marks.

3. Always place trailing periods and commas inside the quotation marks

This is the rule that American English follows (it's different for British English).

Example 3

Excerpt:	The panda could go extinct within the next two decades. Current conservation efforts are not enough.
Sentence:	"The panda could go extinct within the next two decades," Kingsley warns. "Current conservation efforts are not enough."

A comma is required after *decades* to set it apart from the indicator *Kingsley warns*.

4. Put quotation marks around quotes you've incorporated into the context of a sentence

Example 4

Excerpt:	As a chess player, I can say for sure that great chess players have an enhanced ability to recognize patterns.
Sentence:	The author operates under the assumption that "great chess players have an enhanced ability to recognize patterns."

Example 5

Excerpt:	Pandas were once prevalent throughout Asia until infrequent reproduction, poaching, and habitat loss led to thousands of deaths.
Sentence:	Kingsley attributes the rapid decline in the panda population to "infrequent reproduction, poaching, and habitat loss."

Example 6

Excerpt:	The lack of regulation has enabled the destruction of thousands of forests.
Sentence:	Because China takes a hands-off approach to land development, "the lack of regulation has enabled the destruction of thousands of forests."

If the quote includes the beginning of a sentence, you can adjust the capitalization to fit your sentence, as we did in the example above.

5. Semicolons, colons, and dashes that aren't part of the quote should always go outside the quotation marks

Example 7

Excerpt:	We cannot allow our forests to go unprotected.
Sentence:	According to Kingsley, "We cannot allow our forests to go unprotected"; the risks are much too great.

43

6. Question marks and exclamation points belong inside the quotation marks when they're part of the quote, outside when they're not

Example 8

Excerpt:	What would you do if you had a million dollars?
Sentence:	John gets readers thinking from the outset by asking the question, "What would you do if you had a million dollars?"

Example 9

Sentence:	Have you read "The Great Gatsby"?

Example 10

Excerpt:	We must save the environment! Too much is at risk.
Sentence:	"We must save the environment!" Kingsley exclaims.

7. Use single quotation marks for quotes within quotes

Example 11

Excerpt:	Sleep disorders have been linked to diabetes, obesity, cardiovascular disease and depression, and recent research suggests one main cause of "short sleep" is "long light."
Sentence:	Bogard indicates that there are practical reasons too, pointing out that "one main cause of 'short sleep' is 'long light.'"

8. Put quotation marks around the title of the passage

Example 12

Sentence:	In "Bag Ban Bad for Freedom and Environment," Summers describes the advantages plastic bags have over paper bags.

9. Put quotation marks around words referred to as words

Example 13

Excerpt:	It was happening as domestic politics grappled with the merits and consequences of a global war on terror, as a Great Recession was blamed in part on global imbalances in savings, and as world leaders debated a global trade treaty and pacts aimed at addressing climate change.
Sentence:	Goodman uses the word "global" three times in the same sentence to show the extent the world stage influences American issues.

Example 14

Excerpt:	The ragged prisoners crawled out of their cells, desperate for food.
Sentence:	The author uses words such as "ragged," "crawled," and "desperate" to illustrate the bleak conditions of the prison camp.

However, there is no need to put quotation marks around words or short phrases that you aren't calling out as words. In fact, overusing quotation marks can make it seem like you're doubting what the author is saying (scare quotes).

Example 15

Excerpt:	In developed countries, running water is available everywhere and air-conditioning is expected not just in our homes but in our cars. We don't realize that these are luxuries.
Wrong:	The author mentions "running water" and "air-conditioning" as examples of things we take for granted.
Correct:	The author mentions running water and air-conditioning as examples of things we take for granted.

Again, you are allowed to borrow words without having to quote them. Too many students make the mistake of quoting every single word they use from the passage. Words like "running water" and "air-conditioning" are not significant enough by themselves to require quotation marks unless you are referring to them as words as I am in this sentence.

10. You can use an ellipsis to replace words that you want to omit from the middle of a quote.

Example 16

Excerpt:	Researchers have found that pomegranate, which contains a variety of antioxidants, can slow the aging process.
Sentence:	The author cites research that shows that "pomegranate ... can slow the aging process."

However, don't use an ellipsis at the beginning or end of a quote, even if you're omitting the beginning or end of the original sentence.

Example 17

Excerpt:	In fact, compared to paper bags, plastic grocery bags produce fewer greenhouse gas emissions, require 70 percent less energy to make, generate 80 percent less waste, and utilize less than 4 percent of the amount of water needed to manufacture them.
Wrong:	One of the benefits of plastic bags is that they "... produce fewer greenhouse gas emissions..."
Correct:	One of the benefits of plastic bags is that they "produce fewer greenhouse gas emissions."

11. You can use brackets to insert words or letters for clarity or punctuation

Example 18

Excerpt:	Geography is important. It should be taught in every school.
Sentence:	Given the global community we live in today, the author believes that "it [geography] should be taught in every school."

Without the word "geography," the sentence might confuse someone who doesn't know what "it" refers to. We can't assume that everyone has read the original passage.

Example 19

Excerpt:	Reading exposes you to new ideas.
Sentence:	Peters asserts that reading can "expose[] you to new ideas."

In Example 19, we changed the verb tense by removing "s" from "exposes." Empty square brackets indicate this change from the original. In cases where you want to add an "s" to make a verb singular, you can insert a "[s]" after the verb.

Final Note

At the end of the day, proper punctuation is not as important as your actual analysis. A few punctuation errors will not hurt your score. If you're unsure of something, do the best you can and move on.

8

8 Fatal Mistakes Students Make on the Essay

1. Taking a Personal Stance

The SAT essay is an analytical one. While the passage you'll be reading will certainly be opinionated, your essay should not be. You can reiterate what the author's views are in your analysis, but you shouldn't comment on whether the author is right or wrong. The SAT essay is definitely one of those times when you should keep your opinions to yourself.

2. All Summary, No Analysis

One of the easiest traps to fall into is simply summarizing *what* the author's argument is rather than analyzing *how* he or she builds it.

Some students treat the essay almost as an article synopsis in which they just restate what the author says. They completely forget to discuss the techniques the author uses and the impact they have on the reader. This is a huge mistake.

Always make sure your analysis takes center stage. Summary (reiterating) sentences should only be there when you need to add context or clarity to your analysis. Keep them to a minimum.

3. Superficial Analysis

Superficial analysis is stating that something is persuasive but never explaining why specifically. Here are some examples:

- *The statistic that plastic bags are 80 percent more efficient is very persuasive to the reader.*
- *The research done by the National Wildlife Foundation serves as solid evidence. It is an important component of the argument.*
- *The discussion of pollution's harmful effects helps the author's argument.*

Now imagine a full paragraph consisting of statements like the ones above. It would be overly general and vague.

To get a high score, you really need to back up your points and elaborate on them. It's not enough to just call things persuasive.

4. Too Many Quotes

I suspect that this is one of the mistakes that prevented me from getting a 8/8/8 on the May 2016 SAT essay (I got a 7/7/7). My third paragraph was filled with a few too many quotes and for some of them, the quotation marks weren't even necessary according to punctuation rules. Because of these errors, the paragraph came out a bit choppy and I didn't balance it out enough with my own analysis.

While quotes are an essential component of any analytical essay, it goes without saying that your essay should contain more of your own words than someone else's. If you quote too much, your essay will end up reading like a collection of excerpts from the passage.

The solution is to avoid quoting long sentences (direct quotes). If you can remember from the "Narrating & Explaining" chapter, direct quoting isn't the only way to narrate what the author is doing. You can select certain phrases and weave them together or sum things up without any quotes at all. Here are some examples of how you can point something out without using a direct quote:

- The author immediately jumps into a few concrete examples of . . . in order to . . .
- The author finishes his discussion of plastic with a rhetorical question, questioning whether . . .
- He then supplies two pieces of evidence from the National Conservation Society and the World Wildlife Fund.
- Her anecdote about . . . serves to . . .
- The brief background history about . . . helps . . .
- The author cites scientific evidence from . . . showing that . . .

See how the examples above narrate effectively without using a quote? Don't always feel the need to attach a quote to your analysis, especially when the specifics in the quote are unnecessary or when the quote itself is too long and unwieldy to include. For example, don't quote an entire anecdote just to say that the author uses an anecdote. That would be ridiculous. Just say *The author brings up an anecdote about . . . in order to . . .*

A good rule of thumb is that quotes should never make up more than 30% of a paragraph. The meat of your essay should be your analysis, so make sure there's enough of it!

5. Repeating the Same Analysis

Every now and then, a student will write that everything the author does builds trust. The word choice builds trust. The statistics build trust. The appeals to emotion build trust.

Hopefully, you can see why this isn't good.

Even if everything does build trust, present your analysis differently in each paragraph. Explain *how* something builds trust. Talk about the other ways it helps the author's argument. Examine it in a more specific context.

For each point you make, add something new to your analysis. Don't analyze everything in the same way.

6. Too Many Paragraph Markers

This is another mistake that probably cost me a point or two on my essay. I used way too many paragraph markers. A paragraph marker simply specifies the number of the paragraph you're writing about:

In paragraph five, ...

Initially, I thought the use of paragraph markers would be acceptable, even encouraged. After all, there must be a reason paragraph numbers are printed throughout the passage.

But after seeing the scores for my essay and some of my students', my thoughts have changed. Paragraph markers should be avoided. One or two may be fine, but their usage is a sign of lazy writing. Even though they can be used as transitions, they don't add anything to your analysis and having too many will clutter your essay. Furthermore, The College Board has since released several high-scoring sample essays and they rarely if ever use paragraph markers.

7. Not writing fast enough

I'm serious about this one. You should be writing like the wind. You should be writing so fast your hand hurts. If you don't believe me, then you haven't tried writing a full 2.5-3 page SAT essay in 50 minutes.

8. No conclusion

Though missing the conclusion is not as bad as most students think, it's still a glaring error that makes it hard for anyone to ever give you a full score. Always save enough space and enough time for the conclusion. If you miss the conclusion, some graders might not give you the benefit of the doubt—they might just assume you don't know what a conclusion is. A one sentence conclusion is better than no conclusion because at the very least, it shows an understanding of essay organization.

Official Prompts &
Model Essays

At the time of this writing, The College Board has released at least 8 official SAT essay prompts for you to practice with. Most of them can be found at thecollegepanda.com/complete-test-links/

I highly suggest you do at least four practice essays before you sit for the real exam. If you don't practice what you learn, you won't internalize it and you put your score at risk. Just reading this book is not enough.

Here's a partial list of released prompts:

1. "Let There Be Dark" by Paul Bogard

2. "Why Literature Matters" by Dana Gioia

3. "Bag Ban Bad for Freedom and Environment" by Adam B. Summers

4. "Foreign News at a Crisis Point" by Peter S. Goodman

5. Foreword to "Arctic National Wildlife Refuge: Seasons of Life and Land, A Photographic Journey" by Jimmy Carter

6. "Beyond Vietnam—A Time to Break Silence." by Martin Luther King Jr.

7. "The Digital Parent Trap" by Eliana Dockterman

8. "Space Technology: A Critical Investment for Our Nation's Future" by Bobby Braun

This chapter includes model essays to the first four listed. All of them adhere to the template in this book so they're good models to compare your own essays against.

Whereas the next chapter includes perfect-scoring essays written by students, the following essays were written by me. Now why would I include both?

Because any student essay written under pressure in 50 minutes, even if it received a perfect score, is going to have typos, grammar mistakes, inconsistencies, and underdeveloped ideas. While it's great to see that you can make some errors and still get a perfect score, the essays in this chapter make for a better learning experience. They'll give you the clearest picture of what you should be aiming for.

Model Essay to "Let There Be Dark" by Paul Bogard

The idea that we should reduce our electricity usage has its roots in energy conservation but the underlying reasons are often overlooked. In "Let There Be Dark," author Paul Bogard puts forth a detailed argument for the preservation of darkness. In doing so, he employs a variety of literary elements to persuade the reader, including statistics, appeals to authority, and the juxtaposition of past and present.

Bogard's deft use of statistics begins with his discussion of the dark sky. There, he mentions that "8 out of 10 children born in the United States will never know a sky dark enough for the Milky Way." This statistic strikes unsuspecting readers who aren't aware that artificial light overuse is even a problem. Its intent is to not only introduce the issue but also show the extent to which we are losing the night sky. At this point, readers are instilled with a sense of loss, that something they once knew is now going away, but they still may have questions about whether darkness has any real value. Bogard puts those questions to rest by using more statistics. We depend on darkness for the "bats that save American farmers billions in pest control and the moths that pollinate 80% of the world's flora." The implication is that the night sky is just the start of what we'll lose if we continue to let light invade dark. The Earth's ecology relies on darkness and we in turn rely on Earth's ecology for a multitude of benefits. Even skeptical readers who don't appreciate the night sky as much as Bogard does are forced to contend with the practical consequences of the encroachment of day into night. Bogard intensifies the sense of urgency by noting that "the amount of light in the sky increases an average of about 6% every year." The more we ignore the problem of light pollution, the more insurmountable it becomes. Although some may say that Bogard's statistics are overly dramatic, they are grounded in reality and accurately reflect the aftermath in a world without darkness.

Just as persuasive as Bogard's use of statistics is his appeal to authority. He calls upon several credible sources to substantiate his points: "Already the World Health Organization classifies working the night shift as a probable human carcinogen, and the American Medical Association has voiced its unanimous support for 'light pollution reduction efforts and glare reduction efforts at both the national and state levels.'" By showing that there is a consensus among these respected organizations, Bogard gives his argument increased credibility and merit. The mention of the World Health Organization, for example, tells the reader that Bogard's argument is not an arbitrary assault on artificial light but a well-thought-out and researched position. Later, Bogard brings up Paris, the "city of lights." Surely, there is no place that is more a bastion of artificial light pollution than Paris. Why would Bogard mention it? Because "even Paris . . . , which already turns off its monument lighting after 1 a.m., will this summer start to require its shops, offices and public buildings to turn off lights after 2 a.m." The fact that Paris understands the importance of natural darkness puts peer pressure on other cities to do the same. Without these appeals to authority, Bogard's message would lose the credence gained from having others support it.

Lastly, Bogard bolsters his argument by juxtaposing the past and the present. He writes, "All life evolved to the steady rhythm of bright days and dark nights. Today, though, when we feel the closeness of night fall, we reach for a light switch." Here, he paints a sharp contrast between our natural behavior and our behavior today. This contrast continues in Bogard's description of the dark sky in the 1950s and the blanket of light that covers it now. These comparisons of the past and the present highlight the rapid changes that have occurred since light started invading the dark. They convey a sense of foreboding, that the shift has been too sudden for us to cope with and that we're all worse off than before. The benefits of darkness are lost as quickly as we have polluted it with light. "In today's crowded, louder, more fast-paced world, night's darkness can provide solitude, quiet, and stillness," Bogard asserts. By framing darkness as the solution to today's fast-paced world, he uses yet another juxtaposition to persuade. After all, who would prefer crowded and loud to solitude and quiet?

In summary, Bogard—using statistics, appeals to authority, and juxtaposition—makes the case that darkness needs to be maintained and protected. It's his use of persuasive elements that not only inform the reader of the problem but also spur the reader into action.

Model Essay to "Why Literature Matters" by Dana Gioia

The idea that literature plays an integral role in society's development has its roots in academia but its underlying roots are often overlooked. In "Why Literature Matters," Dana Gioia puts forth a detailed argument for the advancement of literature as a bedrock of cultural progress. In doing so, he employs a variety of literary elements to persuade the reader, including statistics, word choice, and quotes from authoritative sources.

Gioia's deft use of statistics begins with his discussion of the arts. He brings up a study that measured the drastic fall in arts participation in America. He specifically draws attention to the fact that the "declines have been most severe among younger adults." By citing these trends, Gioia is grounding his argument in reality so that even skeptical readers won't be able to dismiss it so easily. Not only do the numbers indicate that there is in fact a growing problem but they also show the extent of it. The state of reading in America is in dire straits and because young people participate the least, the troubles will only compound. The underlying assumption that young people are the future of America makes the statistics even more alarming. Gioia further points out that in a poll conducted in 2001, "38 percent of employers complained that local schools inadequately taught reading comprehension." This additional statistic addresses any doubts readers may have over the value of literature engagement. Certainly, literature is important from a cultural standpoint, but Gioia proves that it's also important from a practical standpoint. By covering both bases, he is able to persuade readers with differing perspectives and values. Although Gioia's use of statistics may come off as overly dramatic, it is hard to argue with numbers that paint a bleak picture of an issue that most Americans are familiar with, whether it be in school or in the workplace.

Just as persuasive as his use of statistics is Gioia's word choice. He uses the words "longstanding" and "fundamental" to characterize the cultural activity that is reading literature. Later in the same sentence, he uses the word "deep" to describe the transformations that literature signifies in modern life. His choice of words establishes literature almost as an age-old tradition in America, a piece of cultural heritage that may soon be lost. By appealing to our sense of nostalgia and identity, he hopes to rekindle our connection to and passion for reading. Linking literature to the American identity allows him to cast it as something worth fighting to protect. Indeed, he warns of what will happen if we abandon our literary roots: "As more Americans lose this capacity, our nation becomes less informed, active, and independent-minded." Notice that he explicitly calls out to Americans, again tying his argument to the American values of independence and freedom. Without Gioia's specific choice of words, his argument would lose a strong appeal to our American identities.

Lastly, Gioia bolsters his argument by quoting authoritative sources outside the arts. For instance, he cites "Wired" magazine, which is typically associated more with math and science than with the arts and humanities. Of course, this fact only serves Gioia's argument even more because it implies that even in the cold quantitative world of math, science, and business, literature has its place. He writes, "a new set of mental skills and habits proper to the 21st century [are] decidedly literary in character." He then quotes Daniel Pink, a well-known behavioral science writer, to substantiate his point—"the ability to create artistic and emotional beauty" is one that will be valued more than ever before. Readers of all backgrounds are encouraged to see how relevant literature is outside a humanities domain, that it has applications in our everyday creative and civic spheres. Indeed, Gioia cites studies done by the National Association of Manufacturers and the National Conference of State Legislatures to reinforce this point. The fact that Gioia is able to point to several authorities in a variety of industries that all value an engagement with literature lend great credibility to his main argument.

In summary, Dana Gioia—using statistics, word choice, and authoritative sources—makes the case that we must restore literature back to an active post in our society. It's his use of persuasive elements that not only inform the reader of the problem but also spur the reader into action.

Model Essay to "Bag Ban Bad for Freedom and Environment" by Adam B. Summers

The question of whether to use paper bags or plastic bags has its roots in environmental conservation but its underlying assumptions are often overlooked. In "Bag Ban Bad for Freedom and Environment," author Adam B. Summers puts forth a detailed argument for the use of plastic bags. In doing so, he employs a variety of rhetorical elements to persuade the reader, including an appeal to identity, statistics, and the results of health research.

Summers's deft use of an appeal to identity begins with his discussion of a bill that would ban plastic bags in California. He expresses contempt for "some politicians and environmentalists [who] are now focused on deciding for us what kind of container we can use to carry our groceries." Given the American values of freedom and independence, Summers is aware that this persuasive approach is likely to stir American sensibilities. After all, no one, especially Americans, likes being told what they can and cannot do. Furthermore, Summers frequently uses the collective pronoun "we." This strengthens the unity and sense of purpose of the audience he is addressing and portrays himself as being part of the group. Readers are more likely to accept an argument made by someone who is on their side and aligned with their values. The appeal to identity continues in Summers's definition of a free society: "In a free society, we are able to live our lives as we please, so long as we do not infringe upon the rights of others." By invoking a belief that is so fundamental to American democracy, Summers makes it hard for readers to disagree with him. He frames the right to use a plastic bag as one that all true Americans would rally behind. Although some may say that his appeal to identity is overly dramatic, especially for something as trivial as a plastic bag, it's undoubtedly effective because an American audience understands the larger issues at stake.

Just as persuasive as Summers's appeal to identity is his use of statistics. He quickly challenges the widely-held belief that plastic bags are "evil incarnate" with indisputable numbers: "plastic bags, sacks, and wraps of all kinds …make up only about 1.6 percent of all municipal solid waste materials." In addition, "the most common kind of plastic grocery bags make up 0.3 percent of this total." These small numbers indicate just how inconsequential plastic bags are to the environment. They dispel any limiting assumptions readers may have and open them up to the possibility that plastic bags are not only harmless but also better than paper bags. Indeed, these initial statistics lay the groundwork for even more persuasive ones: "[plastic bags] require 40 percent less energy to make, generate 80 percent less waste, and utilize less than 4 percent of the amount of water needed to manufacture them." By illustrating how much more efficient plastic bags are, Summers persuades the reader from both a cost perspective and an environmental one. It's only through these numbers that he's able to dislodge the widespread idea that paper bags are better. Without them, his message would lose an important distinction between paper and plastic.

Lastly, Summers bolsters his argument by citing the results of scientific research. He brings up several studies showing that "plastic bag bans lead to increased health problems due to food contamination from bacteria that remain in the reusable bags." Furthermore, a statistical analysis by two law professors, Jonathan Klick and Joshua D. Wright, discovered a "spike in hospital emergency room visits" after San Francisco's plastic bag ban was introduced in 2007. The inclusion of this research solidifies the credibility and trustworthiness of Summers's argument. The implication is that if these authoritative sources agree that a plastic bag ban is a bad idea, then we should too. Moreover, Summers has skillfully shown an environmental problem—paper or plastic?—to also be a health problem, something that is far more likely to resonate with readers. Some may not care so much about the environment, but everyone cares about his or her health. Indeed, the specific mention of "E. coli, salmonella, and campylobacter-related intestinal infectious diseases" is sure to elicit a visceral response from many readers.

In summary, Summers—using an appeal to identity, statistics, and health research—effectively makes the case that plastic shopping bags should not be banned. It's his use of persuasive elements that not only inform the reader of the problem but also spur the reader into action.

Model Essay to "Foreign News at a Crisis Point" by Peter S. Goodman

The idea that news should encompass the world, not just America, has its roots in objective journalism practices but its underlying reasons are often overlooked. In "Foreign News at a Crisis Point," writer Peter S. Goodman puts forth a detailed argument that news organizations need to find ways to incorporate foreign affairs into the coverage that Americans consume. In doing so, Goodman employs a variety of literary elements to persuade the reader, including supporting examples, statistics, and strong directives with the pronoun "we."

Goodman's deft use of supporting examples begins with his discussion of the impact of global events on the U.S. He brings up "a global war on terror," "global imbalances in savings," and "a global trade treaty" to show the extent the world stage influences American issues. The word "global" is used three times in the same sentence to highlight this interconnectedness. Additional examples such as the "competition from counterparts on the other side of oceans" create a sense of importance and urgency. Goodman's message to the reader is simple: if we do not understand what is happening in the world at large, we are vulnerable to threats from abroad. Although Goodman's examples may come off as overly dramatic, they are all rooted in issues that Americans contend with on a daily basis. Those who have had to grapple with outsourcing, declining budgets, or terrorist attacks will strongly identify with Goodman's argument.

Just as persuasive as his supporting examples is Goodman's use of statistics. He sets the stage for his argument by noting the decrease in full-time foreign correspondents from 307 to 234. He also points out that the drop in world news coverage is reflected in the 53 percent reduction in newspaper space devoted to foreign affairs. These numbers are likely to surprise readers who aren't aware of the dire state of foreign news, opening their eyes to a real problem that needs to become a priority. They also ground Goodman's argument in reality so that even skeptical readers cannot so easily dismiss his claims. After all, statistics are hard to argue against. Without them, the gravity of the issue might go unrecognized.

Lastly, Goodman bolsters his argument by using strong directives with the pronoun "we." For example, his assertion that we all know the power of social media sets up a baseline assumption from which to further build his argument. The power of social media is a given; it's up to us to use it wisely and creatively to promote foreign news. In laying out his plan for change, Goodman again uses a strong directive with "we": "We need to embrace the present and gear for the future." This is a call to action that readers will find hard to ignore. By using the collective pronoun "we," he establishes a sense of solidarity with the reader, one that extends to society as a whole and the global issues everyone must face together. He frames his cause as one that we must all be a part of and finishes with a rallying cry that urges us to fight for serious-minded journalism: "We need to put back what the Internet has taken away. We need to turn the void into something fresh and compelling." Through these directives, Goodman inspires his audience and empowers them to be more proactive in creating and developing foreign news. He entrusts us with a noble goal and calls on us to be the ones who fulfill it.

In summary, Peter Goodman—using supporting examples, statistics, and strong directives—effectively makes the case that foreign news coverage has declined and must be restored. It's his use of persuasive elements that not only inform the reader of the problem but also spur the reader into action.

10

Real Student Essays that got a Perfect Score

This chapter contains 4 student essays that got perfect 8/8/8 scores. They haven't been edited in any way. I include them for your reference and as proof that my approach works. Not all of them follow the template word for word but all of them showcase concepts I talk about in this book. Screenshots of the actual handwritten essays can be found in Appendix B.

Ashwin Kumar
January 2017 International Exam
"A New Wave of National Parks" by Laura Bush
Uses template? Yes, but with a custom intro and conclusion

The issue of doing more to protect the world's oceans has long been in the hearts and minds of environmentalists, but has not reached a wider audience. Laura Bush attempts to achieve this and spread the message of the need to protect our oceans, using persuasive elements such as the usage of the word "we", statistics, and past precedent.

Bush's adept usage of the word "we" is immediately evident, beginning in the first paragraph. She refers to how waters have "connected us", and urges that "we must intensify our efforts" to conserve our waters. In her usage of collective pronouns, she achieves two goals: she involves herself with the reader, making it clear that it is a shared responsibility, and thus also creates a very effective call to arms on the issue, characterizing it as a problem that must be faced together. By avoiding a lecturing tone and instead using an activist tone—through her use of the word "we"—she can more easily encourage people to take steps to protect the world's oceans. Later, when discussing the dangers of not doing so, she notes that "we are at risk of permanently losing vital marine resources" and that it would harm "our quality of life". In this usage of collective pronouns, she makes the reader feel like the potential losses are his or her losses. Instead of some abstract idea being the motivating factor, readers are instead motivated by personal and hence all the more real risks. In her conclusion, her use

55

of this language continues as she notes that "our wild ocean frontiers are disappearing, and like we did with yellowstone, it is up to us to conserve" the ocean. In this one statement, the reader is now urged to protect his or her "ocean frontier"—hence, a sense of attachment to these vulnerable areas is created, and with this fresh in the reader's mind, once more Bush calls them to action, making it clear that everyone is part of the solution. Without this use of "we", she would lose the activist message and make the subject matter more distant, hampering the argument to protect the world's oceans.

Just as deft as her use of the word "we" is her use of statistics. The reader learns that despite "70% of the world's surface" being covered in ocean, "less than one-half of 1% of the world's oceans are protected in ways that ensure they stay wild". These statistics make the reader think about why such a large part of our world is so vulnerable and unprotected. Bush then writes that "up to 90% of large fish are gone", horrifying the reader, who now understands the consequences of the aforementioned lack of protection. The scale of the problem becomes even more evident as the reader learns of a "plastic garbage dump twice the size of Texas". This statistic is particularly persuasive and impactful as it expresses the magnitude of the issue in the form of a familiar quantity—the large size of Texas. This makes Bush's arguments less abstract and more concrete; it is grounded in facts. Of course, few may fail to see the damage this issue would cause to them—they may argue that fish dying en masse does not matter so long as humans benefit. In a strong, concrete and factual counter, Bush notes that "Nearly half of the world's population lives within 60 miles of the ocean". Now the problem is characterized as a human one, backed by numbers. Some may find Bush's use of statistics to be excessive, but they both ground the argument in reality and also greatly strengthen it against counter arguments, which have to address the data. Thus, the usage of statistics is a persuasive element that greatly contributes to the argument that more should be done to protect the world's oceans.

Finally, Bush makes heavy use of precedent. In talking about Yellowstone, she notes how it was formed at a time when "large wild areas on the frontier were at risk". This project then went on to spark "a 50 year period during which many of the national parks we enjoy today were created". Bush draws an analogy—the circumstances surrounding the creation of Yellowstone is similar to today, with a different frontier at risk. Bush effectively argues that because Yellowstone was so successful, similar modern-day efforts could also be successful. This is especially useful as many readers are likely to agree that Yellowstone was a success and enjoy the park—thus, after reading how she will use it as a "blueprint", they are likely to believe that this analogous project will be successful and will be enjoyed by future generations. Furthermore, showing that this has been done before will help her argue that it can be done now, and makes her argument seem less revolutionary and implausible. For those who believe that creating land-based national parks differs from ocean preservation efforts, Bush then also cites marine monuments like the Papahanaumokuakea Marine National Monument and the Marinas Trench Marine National Monument. With established precedent granting confidence in her plans, her argument is strengthened.'

In conclusion, Bush's clever use of persuasive tools such as statistics, the word "we" and established precedent makes the audience more easily convinced that more should be done to protect the world's oceans.

Jasmine Cheng
January 2017 Exam
"Space Technology: A Critical Investment for Our Nation's Future" by Bobby Braun
Uses template? Yes

The idea of aerospace exploration has its roots in advancing our understanding of the universe, but the underlying factors are often overlooked. In "Space Technology: A Critical Investment for Our Nation's Future," author Bobby Braun outlines an argument for why the US government should continue to invest in NASA. In doing so, he employs a variety of rhetorical elements, including an appeal to the American Identity, strong word choice and real-world applications.

Braun's deft use of appealing to the American Identity begins with his discussion of America's role in the aerospace industry. Braun rallies his readers by indicating that the United States has earned the role of a technological leader, but "to remain the leader ... we must continue to perform research and invest ... preserving a critical component of our nation's economic competitiveness." Braun unites his readers, thus compelling American readers to feel a sense of pride towards their country's accomplishments. In doing so, Braun is able to stress the importance of continuing these achievements in the United States. Braun continues to say that NASA's original goal was "one chosen not for its simplicity, but for its audacity ... to organize and measure the best of our energies and skill." Key concepts such as striving to achieve what may seem impossible and working to the best of your ability are centerpieces of the American dream. The author uses these ideas because they will resonate with any true American patriot. Although some may think that Braun's appeal to the American identity is over the top, it provides an effective technique to stir patriotic attitudes of the reader.

Just as persuasive as Braun's use of patriotic appeal is his strong word choice. The author uses words such as "endangered", "chronic underinvestment" and "stark assessment" to indicate the gravity of the problem. Furthermore, these descriptors are used after a paragraph discussing the past successes of NASA. By setting up a stark contrast between the past greatness of the space program and the bleak present and future it is facing, the author creates a juxtaposition which serves to shock the reader and demonstrate that something needs to be done. Moreover, the author uses his word choice to appeal to the emotions of the reader. His words create a tone of urgency and need for action that is evident throughout the article. Without the author's powerful use of language, the reader would not recognize the significance of what is at stake.

Finally, Braun bolsters his argument with the real world applications of space exploration. The author mentions a multitude of advancements that have resulted from space research: "super computers, solar-and wind-generated energy, the cameras found in many of today's cellphones ... have all benefitted from our nation's investments in aerospace technology." By giving real-world examples of how space research has impacted our everyday lives, Braun implies that continued investment in this result will result in further advancements. Without the use of such examples, some pragmatic readers may question the value of space research—especially for its high cost. However, by stating specific aspects of our lives that have been improved by aerospace research, the author grounds his argument in reality, making it difficult for even skeptical readers to dismiss. In addition to highlighting the applications of space research, Braun also emphasizes its ability to impact future generations through providing jobs. Braun states, "For many of the tens of thousands of engineering and science students in our nation's universities today, the space program provides the opportunity to invent technologies today that will form the foundation for humanity's next great leap across the solar system." Indeed, in an era where competition in the job marketplace is only increasing, the author's indication of the potential jobs fueled by NASA is extremely appealing to readers. Through his use of the real world applications of aerospace technology, the author significantly strengthens his argument that it should be continued.

In summary, Braun—using an appeal to the American identity, strong word choice, and real-world applications—creates powerful argument for why the government must continue to invest in Aerospace technology and research. It is his effective use of persuasive elements that not only inform the readers that there is a problem but also spur the reader into action.

Grace Qing
October 2016 Exam
"The North West Londan Blues" by Zadie Smith
Uses template? No

Obama once said, "A public library is the best manifestation of freedom this nation can offer." This notion of the significance of public libraries is supported by many advocates around the world, many of who see libraries as a niche in the public community. Among them is Zadie Smith. In her article, "The North West London Blues", Smith asserts the need for open public libraries to serve as a haven of cultural prosperity and social development. Accompanied by her utilization of emotion-evoking word choice, logical reasoning, and establishment of ethics, Smith constructs a tailored and logical argument to persuade her audience of this claim.

Smith's use of powerful diction builds a bridge between her and her audience, effectively connecting the two and strengthening her claim. In describing the opposing perspective, she strategically uses emotion evoking vernacular, such as "death", "obsolescence", and "mutilated" to persuade her readers—no one wants to be on the side of "death". These words are accompanied by negative connotations, thus making the reader unconsciously biased away from the opposing argument, and accordingly, to support Smith's. Furthermore, Smith supports her stylistic jargon with persuasive rhetorical, particularly rhetorical questions. By posing questions such as "what kind of a problem is a library?", Smith seems to open it up to interpretation for her audience, and weaken her own claims. However, this actually does the opposite: by giving her audience supposed "freedom of choice", her readers trust her more, and therefore, trust her argument more. Smith intertwines the persuasive strategies of emotional diction and manipulative rhetoric and builds them seamlessly in her argument, allowing her audience to have the allusion of choice, all the while unconsciously feeling a connection with Smith's argument. Her stylistic choices of word choice and rhetorical questions transform her argument to be more than just text on paper, instead building bonds and playing a quintessential role in the persuasiveness of her argument.

Similarly, Smith's use of diction is in line with her objective, logical reasoning throughout her argument, which effectively serves as a fundamental basis of her argument. Smith alludes to the "profits" and "market" to provide a tangible, objective point of view to persuade audience members who are not swayed by her pathos argument. This allows Smith to reach a larger audience, and thus bring power and influence to her claims. Her utilization of facts—"we're humans, not robots"—logically cascades to the numerous functions of public libraries, such as a "public space, access to culture, and preservation of environment". As an audience dedicated to the progress and advancements of society, these all sound like positive traits without bias, thus making the audience question their own prior beliefs of the negatives of open public libraries. This elimination of bias is also supported by Smith's logical concessions that seem to threaten the stability and strength of Smith's claims, but alternatively, strengthens them even further. Smith acknowledges that libraries are not immaculate for society: "money is tight", "there is a hierarchy of needs [in which] public libraries are not top priority", "[libraries] are not hospital beds and classroom size". Smith concedes these points, and by doing so, the audience actually trust her more, and become more prone to take her side. She acknowledges both sides of the debate, thus portraying herself as a knowledgeable, objective advocate that is trying to find the best solution for everyone, not just herself. This eliminates the audience's suspicions that Smith may have her own ulterior motives or subjectivity. Smith's logical reasoning and objective stance add power to her claim and effectively persuade her audience to support her perspective.

Furthermore, Smith's ethical support and morality effectively allow her audience to take her stance on the argument, thus strengthening her persuasiveness. By repeatedly using the word "we", Smith converges the destination between author and reader; she approaches her audience not as a superior or guidance counselor, but as a friend. This connection transforms her argument into one of advice between friends, and she persuades the audience she is just looking out for them. In addition, her epitomization of utilitarianism, the idea of the greatest good for the greatest amount of people, allows the readers to trust her more and her advocacy. She brings into light individuals of all backgrounds—"children", "students", "[the] general public"—to establish herself as a worthy candidate to trust, which makes her audience more inclined to bond and support her claims. Smith's establishment of herself as a moral and ethical figure strengthens her argument to have power and emotion, playing a large role in her persuasiveness.

In hopes of advocating for the need of open public libraries, Smith incorporates persuasive and stylistic elements of emotional jargon, logical reasoning, and establishment as an ethical character. These add power and support her claims, effectively persuading her audience to support the freedom and potential in open public libraries.

Engel Yue
December 2016 Exam
"It's Time For Paid Family and Medical Leave to Empower Working Women and Modernize the Workplace" by Sen. Kirsten Gillibrand
Uses template? No

In a world in which women have just begun to exercise their equal rights, traces of injustice have remained incognito, hidden behind underrated policies of employers and government. Senator Kirsten Gillibrand, in her article, "It's time for Paid Family and Medical Leave to Empower Working Women and Modernize the Workplace," calls for a stand against the traditional "burdens" of women, through the use of relatable imagery, expert opinion, and the feeling of unity.

As a woman, Gillibrand is in a position in which the issue of occupational paid leave is directly relatable. Immediately stating that there is a "stark choice" between a paycheck or taking care of family needs, Gillibrand relies on her diction to convey the desperation that many employees are entangled in a conflict that should not exist. Likewise, her use of the words, "burden" and "disproportionately" draw out the unfair balance that the job system currently has in place. Focusing on women's roles as both mother and "breadmaker" at home and in the office, Gillibrand is able to reach out to women as they can connect to her arguements, able to imagine themselves as being, or one day becoming, the mother who "deserves a day off" to care for a sick child. In regards to her proposal of the Family Act, Gillibrand reiterates the fact that getting this plan across is a "fight". By comparing the struggles of employees in winning job equality and rights to a physical brawl, Gillibrand establishes the notion that this proposal is not of little importance, rather a necessity for the betterment of all employees, and thus, encourages her supporters to battle for "justice".

Moving away from a more emotional appeal, Gillibrand incorporates a variety of expert opinion and statistics to demonstrate the plausibility and high possibility of success for her plan. Gallibrand begins by referencing President Bill Clinton's signing of the Family and Medical Leave Act, twenty years ago. Even though it is implicitly stated, the addition of Clinton's "participation" in this issue of employee paid leave, exemplifies the positive support of Gallibrand's position. Similarly, Gallibrand enlists the words of the current national leader, President Obama, to reassure readers that her Family Act proposal has significant support. Gallibrand builds on the words of Obama's "it's time to do away with workplace policies" to insert her proposal of the Family Act in hopes that readers would be able to form a connection between her plan and the path of the government. In addition, Gallibrand includes statistics to demonstrate that the pros of her proposal outweighs its criticism. As "women now make up almost 50% of the work force", Gallibrand pushes the point that paid leave is now present in a much broader scale in the fields of occupation. When opposers of her plan fear the cost to employers, Gallibrand is able to reassure her supporters and squash the claims by issuing a statistic from a "1998 survey" that revealed "84% of FMLA compliant businesses" reported no losses. Such a significant statistic simply furthers Gallibrand's promise that her similar Family Act will result in a "near-identical" form.

Despite arguement's reliance on facts or strong, supporting imagery, it is the power of Gallibrand's message that unifies her readers and enlists the passion to vote for the Family Act. Calling the era a new "modernized" workplace, Gallibrand sheds the traditional limitations and prejudices. In fact, her proposal includes men's paid leave, as well, covering "paid leave for every U.S. worker." Gallibrand unifies the entirety of the nation as she addresses the "hardworking men and women", as even the name of her proposition is titled, "FAMILY". Gallibrand's ability to connect individuals to their own families, and workers to a united front, allows her to stir up massive agreement through emotional elements of persuasion.

From beginning to end, Gallibrand creates an elegant rhetoric, built on logic, and furthered by appealing to the emotions that readers hold towards their rights, families, and employee status. Focusing not only on one specific group, Senator Kirsten Gallibrand reaches a variety of people in an employee position and is able to provide the promise of compromise between the choice of a paycheck or loved one.

11
Appendix A

BEGIN YOUR ESSAY HERE.

The idea that we should reduce our dependency on fossil fuels has its roots in environmentalism but its underlying reasons are often overlooked. In "Viewpoint: Air-Conditioning Will Be the End of Us," author Eric Klinenberg puts forth a detailed argument that society as a whole must reduce its consumption of air-conditioning to thwart climate change. In doing so, he employs a variety of literary elements and supporting examples, including statistics, concessions, and word choice.

Klinenberg's deft use of statistics begins with his discussion of air-conditioning use in America in paragraph one. He writes, "Today Americans use twice as much energy for air-conditioning as we did 20 years ago," further noting that this is "more than the rest of world's nations combined." This statistic grounds his argument in reality, so that even skeptical readers will not be able to dismiss him so easily. The juxtaposition of America's energy consumption with not only its past but also the rest of the world indicates the intensity of the problem. The implication is that we have drastically increased air-conditioning usage beyond what is reasonable. By using these particular statistics, Klinenberg sounds the alarm on our energy use, signalling to readers just how rampant and excessive their habits of consumption may be. Later, in paragraph 5, Klinenberg brings up yet another statistic — sales of air conditioners rose 20% in India and China last year. Here, he makes it explicit that air conditioning over-use is not a problem distinct to America, but a global one.

Just as persuasive as Klinenberg's use of statistics is his relatability and acknowledgement of the other side. In paragraphs 2

DO NOT WRITE OUTSIDE OF THE BOX.

and 3, he identifies with those who may be hesitant to accept his argument: "I'm hardly against air-conditioning," he asserts. Indeed, he continues to enumerate many advantages of air-conditioning, which include saving "the lives of old, sick and frail people" and enhancing "productivity in offices." By granting these concessions, Klinenberg establishes credibility as someone who has thought through his point of view. Readers are more likely to perceive him not as an extreme environmentalist hell-bent on cutting off the luxuries of AC but as someone who realizes moderate air-conditioning use has its place. Therefore, they are likely to be more receptive to his stances, knowing that their potential objections have been heard. Klinenberg smartly frames himself as an advocate of "public health programs," further creating trust with his audience by putting the greater good first.

Lastly, Klinenberg bolsters his argument with his specific word choice. In paragraph 3, he characterizes the damage caused by air-conditioners as "irreversible." In paragraph 4, he portrays stores in New York City as "pumping arctic air" while "burning through fossil fuels" in "suicidal fashion." By using words such as "arctic" and "suicidal," he invokes an intense kind of imagery, one that dramatizes the excessive way in which we consume A/C. Indeed, we turn our homes, offices, and massive commercial outlets into igloos." Klinenberg's message to the reader is loud and clear: our "burning through fossil fuels" is extreme, superfluous, and even ridiculous. There is simply no need to use up energy at the level we are doing so today.

Although some may say that Klinenberg's choice of words borders on hyperbole, it is an effective way of turning readers' attention to an issue that likely applies to them on a daily basis. It reinforces the point that our own energy habits have left us in a dire situation that will only get worse if changes aren't made.

In summary, writer Eric Klinenberg — using statistics, acknowledgements to the opposing view, and specific word choice — makes the case that rectifying climate change requires a substantial reduction in air-conditioning usage. It's those persuasive elements that not only inform the reader of the problem but also spur the reader into solving it.

12
Appendix B

Ashwin Kumar

Essay - Reading	Essay - Analysis	Essay - Writing
8 2 to 8	**8** 2 to 8	**8** 2 to 8

BEGIN YOUR ESSAY HERE.

The issue of doing more to protect the world's oceans has long been in the hearts and minds of environmentalists, but has not reached a wider audience. Laura Bush attempts to achieve this and spread the message of the need to protect our oceans, using persuasive elements such as usage of the word "we", statistics, and past precedent.

Bush's adept usage of the word "we" is immediately evident, beginning in the first paragraph. She refers to how waters have "connected us", and urges that "we must intensify our efforts" to conserve our waters. In her usage of collective pronouns she achieves two goals: she involves herself with the reader, making it clear that it is a shared responsibility, and thus also creates a very effective call to arms on the issue, characterizing it as a problem that must be tackled together. By avoiding a lecturing tone and intended actions - through her use of "we" - she can more easily encourage people to take steps to protect the world's oceans. Later, when discussing the dangers of not doing so, she notes that "we are at risk of permanently losing vital marine resources" and that it would harm "our quality of life". In this usage of collective pronouns, she makes the reader feel like the potential losses are his or her losses. Instead of some abstract idea being the motivating factor, readers are indeed motivated by personal and hence all the more real risks. In her conclusion, her use of this language continues as she notes that "our wild ocean frontiers are disappearing and like we did with yellowstone, it is up to us to conserve" the ocean. In this one statement, the

DO NOT WRITE OUTSIDE OF THE BOX.

reader now is urged to protect his or her "ocean frontier" - hence, a sense of attachment to these vulnerable areas is created, and with this freshin the reader's mind, once more Bush calls them to action. Making it clear that everyone is part of the solution. Without this use of "we", she would lose the activist message and make the subject matter more distant, hampering the overall argument to protect the world's oceans.

Just as deft as her use of the word "we" is her use of statistics. The reader learns that despite "70% of the world's surface" being covered in ocean, "less than one-half of 1% of the world's oceans are protected" in a way that ensure they stay wild." These statistics make the reader think about why such a large portion of our world is so vulnerable and unprotected. Bush then writes that "up to 90% of large fish are gone", horrifying the reader, who now can understand the consequences of the aforementioned lack of protection. The scale of the problem becomes even more evident as the reader learns of a "plastic garbage dump twice the size of Texas". This statistic is particularly persuasive and impactful as it expresses the magnitude of the issue in the form of a familiar quantity - the large size of Texas. This makes Bush's arguments less abstract and more concrete; it is grounded in facts. Of course, few may see the damage this issue would cause to them - they may argue that fish dying on masse does not matter so long as humans benefit. In a strong, concrete and factual counter, Bush notes that "Nearly half of the world's population lives within 60 miles

DO NOT WRITE OUTSIDE OF THE BOX.

of the ocean". Now the problem is characterized as a human one, backed by numbers. Some may find Bush's use of statistics as excessive, but they both ground the argument in reality and also greatly strengthen it against counter arguments, which have to address the data. Thus, the usage of statistics is a persuasive element that greatly contributes to the argument that more should be done to protect the world's oceans.

Finally, Bush makes heavy use of precedent. In talking about Yellowstone, she notes how it was formed "at a time when large wild areas on the frontier were at risk". This project then went on to start "a 50 year period during which many of the national parks we enjoy today were created". Bush draws an analogy - the circumstances surrounding the creation of Yellowstone is similar to today with a different frontier at risk. Bush effectively argues that because Yellowstone was so successful, similar modern-day efforts could also be successful. This is especially useful as many readers are likely to agree that Yellowstone was a success - thus, after reading about how she sees it as a "blueprint", they are likely to believe that this analogous project will be successful, and will be enjoyed by future generations. Furthermore, showing that it has been done before helps her argue that it can be done now, and makes her argument seem less revolutionary and implausible - rather it is something that has already been done. For those who believe that creating land-based national parks differs from ocean preservation efforts, Bush then also cites marine monuments

DO NOT WRITE OUTSIDE OF THE BOX.

like the Papahanaumokuakea Marine National Monument, and the Marianas Trench Marine National Monument. With established precedent granting confidence in her plan, her argument is strengthened.

In conclusion, Bush's clever use of persuasive tools such as statistics, the word "well and established precedent makes the audience more easily convinced that more should be done to protect the world's oceans.

Jasmine Cheng

Essay Scores

Essay - Reading	Essay - Analysis	Essay - Writing
8 2 to 8	**8** 2 to 8	**8** 2 to 8

Essay Details

BEGIN YOUR ESSAY HERE.

The idea of aerospace exploration has its roots in advancing our understanding of the universe, but the underlying factors are often overlooked. In "Space Technology: A Critical Investment for Our Nation's Future", author Bobby Braun outlines an argument for why the US government should continue to invest in NASA. In doing so, he employs a variety of rhetorical elements, including an appeal to the American Identity, strong word choice and real-world applications.

Braun's deft use of appealing to the American Identity begins with his discussion of America's role in the aerospace industry. Braun rallies his readers by indicating that the United States has earned the role of a technological leader, but "to remain the leader, we must continue to perform research and invest ... preserving a critical component of our nation's economic competitiveness." Braun unites his readers, thus compelling American readers to feel a sense of pride towards their country's accomplishments. In doing so, Braun is able to stress the importance of continuing these achievements in the United States. Braun continues to say that NASA's original goal was "one chosen not for its simplicity, but for its audacity ... to "organize and measure the best of our energies and skill." Key concepts such as striving to achieve what may seem impossible and working to the best of your ability are centerpieces of the American

DO NOT WRITE OUTSIDE OF THE BOX.

dream. The author uses these ideas because they will resonate with any true American patriot. Although some may think that Braun's appeal to the American identity is over the top, it provides an effective technique to stir patriotic attitudes of the reader.

Just as persuasive as Braun's use of patriotic appeal is his strong word choice. The author uses words such as "endangered", "chronic underinvestment" and "stark assessment" to indicate the gravity of the problem. Furthermore, these descriptors are used after a paragraph discussing the past successes of NASA. By setting up a stark contrast between the past greatness of the space program and the bleak present and future it is facing, the author creates a juxtaposition which serves to shock the reader and demonstrate that something needs to be done. Moreover, the author uses his word choice to appeal to the emotions of the reader. His words create a tone of urgency and need for action that is echoed throughout the article. Without the author's powerful use of language, the reader would not recognize the significance of what is at stake.

Finally, Braun bolsters his argument with the real world applications of space exploration. The author mentions a multitude of advancements that have resulted from space research:

DO NOT WRITE OUTSIDE OF THE BOX.

"Super computers, solar and wind generated energy, the cameras found in many of today's cellphones ... have all benefitted from our nation's investments in aerospace technology." By giving real-world examples of how space research has impacted our everyday lives, Braun implies that continued investment in this result will result in further advancements. Without the use of our examples, some pragmatic readers may question the value of space research — especially for its high cost. However, by stating specific aspects of our lives that have been impacted by aerospace research, the author grounds his argument in reality, making it difficult for even skeptical readers to dismiss. In addition to highlighting the application of space research, Braun also emphasizes its ability to impact future generations through providing jobs. Braun states, "for many of the tens of thousands of engineering and science students in our nation's universities today, the space program provides the opportunity to invest technologies today that will form the foundation for humanity's next great leap across the solar system." Indeed, in an era where competition in the job marketplace is only increasing, the author's indication of the potential jobs fueled by NASA is extremely appealing to readers. Through his use of the "real world

DO NOT WRITE OUTSIDE OF THE BOX.

applications of aerospace technology, the author significantly strengthens his argument that it should be continued.

In summary, Braun — using an appeal to the American identity, strong word choice, and real-world applications — creates powerful argument for why the government must continue to invest in Aerospace technology and research. It is his effective use of persuasive elements that not only inform the readers that there is a problem but also spur the reader into action.

Grace Qing

Essay - Reading	Essay - Analysis	Essay - Writing
8	8	8

Essay Skills Insight

Obama once said, "A public library is the best monu-ument statue of freedom this nation can offer." This notion of the significance of public libraries is supported by many advocates around the world, many of who see libraries as a niche in the public community. Among them is Zadie Smith. In her article, "The North West London Blues," Smith asserts the need for open public libraries to serve as a haven of cultural prosperity and social development. Accompanied by her utilization of emotion-evoking word choice, logical reasoning, and establishment of ethics, Smith constructs a tailored and logical argument to persuade her audience of this claim.

Smith's use of powerful diction builds a bridge between her and her audience, effectively connecting the two and strengthening her claim. In describing the opposing perspective, she strategically uses emotion-evoking vernacular, such as "death," "obsolescence," and "mutilated" to persuade her readers — no one wants to be on the side of "death". These words are accompanied by negative connotations, thus making the reader unconsciously bias away from the opposing argument, and accordingly, to support Smith's. Furthermore, Smith supports her stylistic jargon with persuasive rhetorical, particularly rhetorical questions. By posing questions such as "what kind of a problem is a library?", Smith seems to open it up to interpretation for her audience, and weaken her

own claims. However, this actually does the opposite: by giving her audience supposed "freedom of choice," her readers trust her more, and therefore, trust her argument more. Smith intertwines the persuasive strategies of emotional diction and manipulative rhetoric, and builds them seamlessly in her argument, allowing her audience to have the allusion of choice, all the while unconsciously feeling a connection with Smith's argument. Her stylistic choices of word choice and rhetorical questions transform her argument to be more than just text on paper, instead building bonds and playing a quintessential role in the persuasiveness of her argument.

Similarly, Smith's use of diction is in line with her objective, logical reasoning throughout her argument, which effectively serves as a fundamental basis of her argument. Smith alludes to the "profits" and "market" to provide a tangible, objective point of view to persuade audience members who are not swayed by her pathos argument. This allows Smith to reach a larger audience, and thus bring power and influence to her claims. Her utilization of facts — "we're humans, not robots" — logically cascades to the numerous functions of public libraries, such as a "public space, access to culture, and preservation of environment". As an audience

dedicated to the progress and advancements of society, these all sound like positive traits without bias, thus making the audience question their own prior beliefs of the negatives of open public libraries. This elimination of bias is also supported by Smith's logical concessions that seem to threaten the stability and strength of Smith's claims, but alternatively, strengthens them even further. Smith acknowledges that libraries are not all immaculate for society: "money is tight", there is a hierarchy of needs [in which] public libraries are not top priority", "[libraries] are not hospital beds and classroom size". Smith concedes these points, and by doing so, the audience actually trust her more, and become more prone to take her side. She acknowledges both sides of the debate, thus portraying herself as a knowledgable, objective advocate that is trying to find the best solution for everyone, not just herself. This eliminates the audience's suspicions that Smith may have her own ulterior motives or subjectivity. Smith's logical reasoning and objective stance add power to her claim and effectively persuade her audience to support her perspective.

Furthermore, Smith's ethical support and morality effectively allow her audience to take her stance on the argument, thus strengthening her persuasiveness.

By repeatedly using the word "we", Smith converges the distinction between author and reader. She approaches her audience not as a superior or guidance counselor but as a friend. This connection transforms her argument into one of advice between friends, and she persuades the audience she is just looking out for them. In addition, her epitomization of utilitarianism, the idea of the greatest good for the greatest amount of people, allows the readers to trust her more and her advocacy. She brings into light individuals of all backgrounds — "children", "students", [the] general public" — to establish herself as a worthy candidate to trust, which makes her audience more inclined to bond and support her claims. Smith's establishment of herself as a moral and ethical figure strengthens her arguments to have power and emotion, playing a large role in her persuasiveness.

In hopes of advocating for the need of open public libraries, Smith incorporates persuasive and stylistic elements of emotional jargon, logical reasoning, and establishment as an ethical character. These add power and support her claims, effectively persuading her audience to support the freedom and potential in open public libraries.

Engel Yue

Essay - Reading	Essay - Analysis	Essay - Writing
8 2 to 8	**8** 2 to 8	**8** 2 to 8

BEGIN YOUR ESSAY HERE.

In a world in which women have just begun to exercise their equal rights, traces of injustice have remained incognito, hidden behind underrated policies of employers and government. Senator Kirsten Gillibrand, in her article, "It's Time for Paid Family and Medical Leave to Empower Working Women and Modernize the Workplace", calls for a stand against the traditional "burdens" of women, through the use of relatable imagery, expert opinion, and the feeling of unity.

As a woman, Gillibrand is in a position in which the issue of occupational paid leave is directly relatable. Immediately stating that there is a "stark choice" between a paycheck or taking care of family needs, Gillibrand relies on her diction to convey the desperation that many employees are entangled in a conflict that should not exist. Likewise, her use of the words, "burden" and "disproportionately" draw out the unfair balance that the job system currently has in place. Focusing on women's roles as both mother and "breadmaker" at home and in the office, Gillibrand is able to reach out to women as they can connect to her arguments, able to imagine themselves as being, or one day becoming, the mother who "deserves a day off" to care for a sick child. In regards to her proposal of the FAMILY Act, Gillibrand reiterates the fact that getting this plan across is a "fight". By comparing the struggles of employees in winning job equality and rights to a physical brawl, Gillibrand establishes the notion that this proposal is not of little importance, rather a necessity for the

DO NOT WRITE OUTSIDE OF THE BOX.

betterment of all employees, and thus, encourages her supporters to battle for "justice".

Moving away from a more emotional appeal, Gillibrand incorporates a variety of expert opinion and statistics to demonstrate the plausibility and high possibility of success for her plan. Gillibrand begins by referencing President Bill Clinton's signing of the Family and Medical Leave Act, twenty one years ago. Even though it is implicitly stated, the addition of Clinton's "participation" in this issue of employer paid leave, exemplifies the positive support of Gillibrand's position. Similarly, Gillibrand enlists the words of the current national leader, President Obama, to reassure readers that her FAMILY Act proposal has significant support. Gillibrand builds on the words of Obama's "it's time to do away with workplace policies" to insert her proposal of the FAMILY Act in hopes that readers would be able to form a connection between her plan and the path of the government. In addition, Gillibrand includes statistics to demonstrate that the pros of her proposal outweighs its criticism. As "women now make up almost 50% of the work force", Gillibrand pushes the point that paid leave is now present in a much broader scale in the fields of occupation. When opposers of her plan fear the cost to employers, Gillibrand is able to reassure her supporters and squash the claims by issuing a statistic from

DO NOT WRITE OUTSIDE OF THE BOX.

a "1998 survey" that revealed "84% of FMLA compliant businesses" reported no losses. Such a significant statistic simply furthers Gillibrand's promise that her similar FAMILY Act will result in a "near-identical" form.

Despite argument's reliance on facts or strong, supporting imagery, it is the power of Gillibrand's message that unifies her readers and enlists the passion to vote for the FAMILY Act. Calling the era a new "modernized" workplace, Gillibrand sheds the traditional limitations and prejudices. In fact, her proposal includes men's paid leave, as well, covering "paid leave for every U.S Worker." Gillibrand unifies the entirety of the nation as she addresses the "hardworking men and women", as even the name of her proposition is titled, "FAMILY". Gillibrand's ability to connect individuals to their own families, and workers to a united front, allows her to stir up massive agreement through emotional elements of persuasion.

From beginning to end, Gillibrand creates an elegant rhetoric, built on logic, and furthered by appealing to the emotions that readers hold towards their rights, families, and employee status. Focusing not only on one specific group, Senator Kirsten Gillibrand reaches a variety of people in an employee position and is able to provide the promise of compromise between the choice of a paycheck or loved one.